T0331087

MULTIMEDIA COMPUTING SYSTEMS AND VIRTUAL REALITY

MULTIMEDIA COMPUTING SYSTEMS AND VIRTUAL REALITY

Edited by
Rajeev Tiwari, Neelam Duhan,
Mamta Mittal, Abhineet Anand and
Muhammad Attique Khan

CRC Press
Taylor & Francis Group
Boca Raton London New York

CRC Press is an imprint of the
Taylor & Francis Group, an **informa** business

First edition published 2022
by CRC Press
4 Park Square, Milton Park, Abingdon, Oxon, OX14 4RN

and by CRC Press
6000 Broken Sound Parkway NW, Suite 300, Boca Raton, FL 33487-2742

CRC Press is an imprint of Informa UK Limited

Library of Congress Cataloging-in-Publication Data
A catalogue record for this book is available from the British Library

Names: Tiwari, Rajeev, editor.
Title: Multimedia computing systems and virtual reality / edited by Rajeev Tiwari, Neelam Duhan, Mamta Mittal, Abhineet Anand, Muhammad Attique Khan.
Description: First edition. | Abingdon, Oxon; Boca Raton, FL : CRC Press, [2022] | Includes bibliographical references and index. |
Summary: "This book focuses on futuristic trends and innovations in multimedia systems using big data, IoT and cloud technologies. The authors present recent advancements in multimedia systems as they relate to various application areas such as health care services and agriculture-related industries"-- Provided by publisher.
Identifiers: LCCN 2021044785 (print) | LCCN 2021044786 (ebook) |
ISBN 9781032048239 (hbk) | ISBN 9781032052335 (pbk) | ISBN 9781003196686 (ebk)
Subjects: LCSH: Multimedia systems. | Virtual reality.
Classification: LCC QA76.575 .M79124 2022 (print) | LCC QA76.575 (ebook) | DDC 006.7--dc23/eng/20211105
LC record available at https://lccn.loc.gov/2021044785
LC ebook record available at https://lccn.loc.gov/2021044786

ISBN: 978-1-032-04823-9 (hbk)
ISBN: 978-1-032-05233-5 (pbk)
ISBN: 978-1-003-19668-6 (ebk)

DOI: 10.1201/9781003196686

Typeset in Times
by MPS Limited, Dehradun

Contents

Editors

Dr. Rajeev Tiwari is a senior associate professor in SCS in UPES, Dehradun, India. He is a senior IEEE Member and has more than 15 years of research and teaching experience. His broad areas of research includes cloud computing, MANET, VANET, fog computing, QoS in such network environments, cache invalidation techniques, Internet of Things (IoT), big data analytics, and machine learning.

Dr. Neelam Duhan is currently working as an associate professor in the Department of Computer Engineering at J. C. Bose University of Science and Technology, YMCA, Faridabad, India. She received her B.Tech. in computer science and engineering, M.Tech. in computer engineering and Ph.D. in computer engineering in 2002, 2005 and 2011, respectively. She has successfully guided three Ph.Ds and is currently guiding four Ph.D. scholars in the areas of machine learning, semantic web, route prediction, and social networks.

Dr. Mamta Mittal graduated in computer science and engineering from Kurukshetra University Kurukshetra, India in 2001 and received a master's degree (Honors) in computer science and engineering from the YMCA, Faridabad. She completed her Ph.D. from Thapar University, Patiala, India in computer science and engineering. Dr. Mittal has 18 years of teaching experience and her research areas include data mining, big data, machine learning, soft computing, and data structure. She is also a lifetime member of the CSI.

Dr. Abhineet Anand is a professor in the computer science and engineering department with Chitkara University, Punjab, India. He has 20+ years of academic and administrative experience, and his research includes cloud computing, cloud security, decision tree, nearest neighbor method, clustering, rule induction, optical fiber wwitching in wavelength multiplexing, and automata theory.

Dr. Muhammad Attique Khan earned his master's degree in human activity recognition for application of video surveillance from the COMSATS University Islamabad, Pakistan. Currently, he is doing his Ph.D. from the COMSATS University Islamabad, Pakistan. He is currently a lecturer in the computer science department in HITEC University Taxila, Pakistan. His primary research focui in recent years is medical imaging, MRI analysis, video surveillance, human gait recognition, and agriculture plants.

Contributors

Ahad Abro
Department of Informatics
Ege University
Izmir, Turkey

Harjeev Singh Ahluwalia
Value Labs LLP
Hyderabad, Telangana, India

Abhineet Anand
Chitkara University Institute of
Engineering and Technology
Chitkara University
Punjab, India

Komal Kumar Bhatia
J.C. Bose University of Science and
Technology, YMCA
Faridabad, Haryana, India

P. Deepan
Annamalai University
Annamalainagar, Chidambaram,
Tamilnadu, India

Neelam Duhan
J.C. Bose University of Science and
Technology, YMCA
Faridabad, Haryana, India

G S Pradeep Ghantasala
Chitkara University Institute of
Engineering and Technology
Chitkara University
Punjab, India

H.M.K.K.M.B. Herath
Department of Mechanical Engineering,
Faculty of Engineering Technology
The Open University of Sri Lanka
Nugegoda, Sri Lanka

Awais Khan Jumani
Department of Computer Science
ILMA University
Karachi, Sindh, Pakistan

Isha Kansal
Chitkara University Institute of
Engineering and Technology
Chitkara University
Punjab, India

Abdullah Ayub Khan
Department of Computer Science
Sindh Madressatul Islam University
Karachi, Sindh, Pakistan

Asif Ali Laghari
Department of Computer Science
Sindh Madressatul Islam University
Karachi, Sindh, Pakistan

Umesh Kumar Lilhore
Chitkara University Institute of
Engineering and Technology
Chitkara University
Punjab, India

B. G. D. A. Madhusanka
School of Science and Engineering
Malaysia University of Science and
Technology (MUST)
Petaling Jaya, Malaysia

Rishabh Nanawati
Computer Engineering, Mukesh Patel
School of Technology, Management
& Engineering
NMIMS University
Mumbai, India

Pulkit Narwal
J.C. Bose University of Science and
Technology, YMCA
Faridabad, Haryana, India

Renu Popli
Chitkara University Institute of
Engineering and Technology
Chitkara University
Punjab, India

T. Veni Priya
Annamalai University, Annamalainagar,
Chidambaram
Tamilnadu, India

Premkumar Rajagopal
School of Science and Engineering
Malaysia University of Science and
Technology (MUST)
Petaling Jaya, Malaysia

Sureswaran Ramadass
School of Science and Engineering
Malaysia University of Science and
Technology (MUST)
Petaling Jaya, Malaysia

Aman Sariya
Computer Engineering, Mukesh Patel
School of Technology, Management
& Engineering
NMIMS University
Mumbai, India

Kaustubhi Shuklaa
ImaginXP
Pune, Maharastra, India

Waqas Ahmed Siddique
Department of Computer Science
ILMA University
Karachi, Sindh, Pakistan

Sarita Simaiya
Chitkara University Institute of
Engineering and Technology
Chitkara University
Punjab, India

Sheetal Soni
Assistant Professor, National Institute
of Fashion Technology
Jodhpur, Rajasthan, India

L.R. Sudha
Annamalai University, Annamalainagar,
Chidambaram
Tamilnadu, India

Rajeev Tiwari
School of Computer Science
University of Petroleum & Energy Studies
Dehradun, India

Naresh Kumar Tiwari
Chitkara University Institute of
Engineering and Technology
Chitkara University
Punjab, India

R. Raja Vignesh
K.S.K College of Engineering and
Technology
Kumbakonam, Tamilnadu, India

Usha Yadav
National Institute of Fashion
Technology
Jodhpur, Rajasthan, India

1 Use of Virtual Reality in Exposure Therapy and Other Psychological Treatment Methods

*Aman Sariya, Rishabh Nanawati
and Supriya Agarwal*
Department of Computer Engineering, Mukesh Patel School
of Technology, Management & Engineering, NMIMS
University, Mumbai, India

1.1 INTRODUCTION

Virtual reality (VR) is a well-known technology that has seen a surge of growth across multiple fields like gaming, marketing, customer service training and education. In the field of medical science, it is known to be used in medical training for a deeper understanding and research of anatomy and surgery. It is also used by doctors and surgeons in diagnosis and surgery to visualise various body parts of a patient, developed from scanning techniques like MRIs and CT scans [1]. Its applications in the field of healthcare are expanding universally and gaining traction as its potential is being recognised.

Although there are many applications of VR in physical healthcare, it has proved to benefit one's mental well-being as well. Several VR applications have been developed in the last two decades for the understanding, assessment and treatment of mental health disorders [2]. Many studies have been conducted involving trials of these applications on patients with various disorders. Results show that participants are already being affected by an array of virtual environments, fully knowing they are not real [3].

While early development of VR dates back to the 1950s, psychologists did not start using VR as a part of their treatment until the 1990s. The first study to examine the use of VR in exposure therapy investigated its effects on the treatment of acrophobia (fear of heights) and positive results prompted further research in the field [4]. Thus far, VR has been studied as a tool in the treatment of many psychiatric disorders like anxiety disorders, stress-related disorders, psychosis and eating disorders.

In terms of economic worth, VR in healthcare was valued at USD$2.14 billion as of 2019 and is estimated to grow to USD$33.72 billion by 2027 on the continent of North America alone. The share of mental-health applications in this valuation was 33.4% [5].

DOI: 10.1201/9781003196686-1

In the chapter, various studies that highlight the different applications of VR in the field of mental health have been reviewed. To provide the context that is needed to understand these applications, it begins with an introduction of the psychological science behind the applications. This is followed by an explanation of the working of a VR simulation and how it manages to create a virtual near-real environment. After highlighting the features and benefits of such a VR application, it focuses on different types of mental health disorders and the effectiveness of VR based treatments on them. Additionally, it then explores various issues associated with the use of such an application, which is followed by a discussion where the authors provide a look into their own research efforts and the future scope of this technology.

1.2 PSYCHOLOGY

Stimuli are events or objects that induce a particular chain of thoughts. Stimuli drive feelings like happiness, sadness and fear. Steinman et al. [6] define exposure therapy as *"any treatment that encourages the systematic confrontation of stimuli that are feared, with the aim of decreasing fearful physiological, cognitive, and behavioural reactions."* These feared stimuli can be external to the patient, like real-life circumstances or environmental objects on which the patient has no control, or entities internal to the patient, like feared notions or sensations of which the patient has some control, if not all.

Here is an example of how a therapist helps its patient overcome their atrocities. A child has been nagging their parent to teach them how to ride a bicycle. The parent follows a classical approach – they ask their child to get on the bicycle and assures them that they will hold and support it. The child is fearful because they do not believe they will be able to maintain balance on the bicycle. They start slowly and the child starts feeling confident because they believe the parent is supporting the bicycle's balance. In the next few runs, the parent supports the bike inter-mittently, without the child realising when the parent leaves and catches on again. The realisation between the parent supporting and not supporting the bike is blurred and unrecognisable to the child. From the child's perspective, the cycle is being supported constantly. This makes their mind forget about its actual fear and focuses more on moving forward. The child, constantly facing the phobia, is slowly getting desensitised to it and starts gaining trust in their own abilities. For the final run, the parent may just hold the bike initially and leave as soon as the child starts peddling. By now, the child has gained momentum, confidence and trust in their abilities.

Similarly, the therapist plays the role of the parent, providing the support me-chanism needed to the child, who is the patient. The fear is the loss of balance leading to the child falling down, and the environment corresponds to the cycle, the support and the road.

In exposure therapy, clients are first subjected to lower intensity stimuli (at the bottom of the exposure hierarchy) repeatedly until that fear abates. They then move on to the higher-intensity stimuli. This process is repeated until no stimuli in the process elicits fear any more [7]. This level-wise graded approach usually involves construction of an exposure hierarchy by the mental health professional and the patient, where stimuli are ranked based on how fearful the patient is of various

TABLE 1.1

Example of Exposure Hierarchy for the Fear of Public Speaking

Difficulty Level	Scenario
1	Patient stands in an empty room.
2	Patient observes a group of people having a discussion from a large distance.
3	Patient observes the same group of people from a shorter distance.
4	Patient becomes a part of the same group and observes the participants but does not engage in conversation.
5	Patient is now engaged in small talk with the characters in the simulation with one-word answers.
6	Patient is placed in a similar environment but with more people with question prompts requiring longer answers.
7	Patient is placed on stage in an empty auditorium with spotlights on them. They may be required to read out a small paragraph to serve as a speech.
8	Patient is placed in a full-capacity auditorium with spotlights and is required to read out a prompt or personal speech to which the audience may applaud.

scenarios. The client is then subjected to a series of scenarios that trigger their fears. The scenarios increase in intensity based on the progress of the patient through the course of treatment. A commonly used scale for these rankings is the "Subjective Units of Discomfort (SUD)" scale [8]. It assigns values 0–100 to measure the intensity of the disturbance that the patient is feeling. The pathological fear that is to be made redundant, is modified by integrating corrective information with the same fear structure through exposure therapy [9] (Table 1.1).

The two common approaches for clinical delivery of exposure therapy are imaginal exposure and in-vivo exposure [10].

Imaginal exposure (IE) [11] involves encouraging the patient to imagine their feared scenarios or traumatic events. A therapist sits the client down on a comfortable chair and asks them to close their eyes while performing relaxation exercises. The client is then made to imagine a scenario as though they are in a movie. Scenarios, based on the discomfort level, can be very elaborate. They involve descriptions of sensory elements like taste, touch, smell, sound and vision along with descriptions of the settings, the people, if any, around the client as well as his or her actions. For example, for social anxiety, the client may be made to imagine that they are entering a coffee shop. Their actions might be picking a table, ordering coffee at the counter, encountering other customers, waiting in line for the washroom, etc. They can be made to imagine the taste and aroma of the coffee, what they see around them, the music playing, the texture of the coffee mug, etc. These intricacies generate a sense of immersion and introduce a sense of realism for the client. The therapist may continuously gauge the reaction of the client. After the therapist concludes the description of the scenario, they ask the client a series of questions about what they see, hear, think, feel emotionally or physiologically, etc. Some therapists may choose to record these sessions for clients to listen to them as homework and rate their level of discomfort. This serves as a benchmark to measure progress of the client through the course of the treatment.

The second, perhaps a more effective method to deliver exposure therapy, is the in-vivo approach. This approach involves subjecting the client to the actual fear-evoking stimuli. Here, the therapist is present with the client when they experience thoughts of fear and anxiety. The therapist guides them towards navigating these thoughts and dealing with them in the best way possible. Slowly, the need for the therapist's guidance diminishes and the patient becomes independent enough to be able to deal with their fears. In vivo is a more effective method for the simple reason that it has the potential to stimulate all your senses to provide a sense of authenticity.

Let us take a look at how virtual reality fits in exposure therapy. While imaginal exposure has proven to be an effective way to begin therapy, it tends to be very slow unless supported by in-vivo exposure because it depends on the capabilities of the client to evoke feared imagery in depth [7,12]. This may be an added stressor for the patient, as they already feel overwhelmed by their feelings of anxiety and imagining the situation that caused it may not help. Assuming that they do imagine it, their overburdened mind might not be able to imagine all the details accurately or for the amount of time required for it to be effective. It also does not stimulate the patient's senses enough to have a solidified long-lasting effect.

In vivo, on the other hand, is a very effective way to deliver exposure therapy because it involves the client actually being present in the situation. However, it does have multiple limitations. The client may be reluctant to participate in such a form of therapy because the scenarios could be unsettling and gravely anxiety-inducing. This leads to higher chances of abandonment of the treatment. Further, it is not easy to set up in-vivo scenarios. It may either be too costly, unrealistic, difficult or even impossible. If the client has a fear of flying, it is very difficult to simulate scenarios to tackle specific fears like take-off, landing or turbulence. Similarly, for other fears involving unpredictable events like a natural disaster or an existential fear like the fear of losing a loved one, in vivo might not be of much use. Hence, the scope of fears that in vivo can deal with is limited.

These are the gaps in imaginal and in-vivo exposure therapy that virtual reality has the potential to bridge. VR–based exposure therapy (VRET) has the capability to stimulate one's sense of vision, hearing, touch, and with the correct clinical environment and delivery method, even smell and taste by replacing real anxiety-evoking scenarios with their virtual counterparts. This may help to encourage people who would be disinclined to confront high intensity stimuli to seek help, ultimately aiding in therapeutic habituation (i.e., a behavioural response decrement that results from repeated stimulation and that does not involve sensory adaptation [13]). There is a plethora of research, randomised controlled trials (RCTs) and meta-analysis that have been conducted that strengthen the proof of the efficacy of VRET.

1.3 THE WORKINGS AND QUALITY OF THE SIMULATION

In order to work the VR simulation, the user must be equipped with a head-mounted display (HMD) and a software control system (which may be embedded in the HMD), along with other optional input devices (like data gloves, motion controllers, etc.) and output devices (like in-ear earphones, haptic gloves, etc.). The user's movements are fed to the software as input via the input devices. The software performs various functions based on those inputs and produces outputs which are fed back in the form of graphic changes, auditory output, etc [14].

VR is viewed in medicine in one of two ways: VR as a simulation tool and VR as an interaction tool. For its use in psychotherapy, VR is more prevalent as a simulation tool because it provides a lifelike human-computer interface paradigm that eliminates the need for the user to be a third-party observer disconnected from the virtual surroundings. Instead, it allows the user to be an active participant in a virtual world.

1.3.1 INTERVENTION APPROACH

VRET may be employed in different ways to treat a particular disorder. At the foundation of these approaches lies the carefully designed virtual environment, consisting of both interactive and non-interactive objects. The traditional approach is to have the patient use the simulation in a clinic, where their therapist is able to see what the patient sees, giving them tasks to perform and guiding them throughout the process.

A new, rather modern approach, is to employ an automated and gamified simulation that would use tasks, that rely on game mechanics, and points, badges and trophies, that reinforce progress and achievements as the patient completes various tasks and progress with their conditions [15]. Modern games are quite immersive, and their elements like level design, story and narrative building, feedback of performance and avatar assistance can be very helpful in the treatment of psychiatric disorders. Moreover, these characteristics would make the simulation guided by itself and would eliminate the need of a therapist's presence. However, the psychiatric community still remains apprehensive about a gamified, automated simulation being delivered in the absence of a mental health professional. There are concerns of an automated simulation misguiding a patient, especially if the patient is not aware of his condition enough and cannot tell if something is right for them. There is also the added worry about the absence of a professional to handle the patient if the patient's condition worsens or if they get triggered, causing undesirable reactions like episodes of panic attacks.

Hence, most VRET interventions employ a mixed approach. The simulation is gamified and assists patients wherever necessary, while being administered in the presence of a therapist. The therapist monitors the patient's actions, observes their reactions and is available to guide them throughout. This would make the treatment more interactive and immersive while remaining relatively safer.

1.3.2 EFFICACY

In order to make the most effective use of VR simulations, developers must build a simulation that "tricks" the user's senses into believing that the virtual world is real [16]. Resource allocations and utilisation needs to be done carefully by the developer in this case. Which senses to stimulate, how much stimulation is necessary, what quality of stimulation to achieve for the user to perceive them as real – these are a few questions that every developer needs to answer before beginning with the development.

VR simulations can be employed in a number of ways, ranging from the use of only a head-mounted display (HMD) to full-body interaction with virtual object through motion capture. The user's perceptual experience depends on the *level of immersion*, which is a technical concept that can be applied to various intervention scenarios [17]. Level of immersion has been known to depend mainly on five aspects – inclusiveness, extensiveness, surroundings, vividness and matching.

- *Inclusivity* or *inclusiveness* relates to the ability of a VR simulation to eradicate indicators that remind the user of a physical world. This includes the role of devices like joysticks, noise from external sources, weight of instruments worn, etc.
- *Extensiveness* is the number of sensory modalities that have been manipulated.
- *Surrounding* refers to the appearance of the VR technology, which includes the head-mounted display and/or the computer screen.
- *Vividness* involves the resolution and accuracy of the depth and colour that is presented in the virtual environment and how close it is to reality.
- *Matching* refers to the field of view of the user, the various changes that occur in it and how synchronized they feel with the user's interactions.

Another component that plays an important role in developing VR for clinical use is the *sense of presence*. It is the perceptual experience of the user which is understood by how engaged the user feels in the virtual environment. It is facilitated by the user's perception thresholds and the tasks performed by them [18].

The two concepts are distinctly based on their method of determination. While level of immersion is a technical concept that is derived from the VR simulation, sense of presence is based on each individual's experience with the simulation. However, it is not possible to completely dissociate them, as sense of presence is facilitated by the level of immersion. Simulations with high levels of immersion are often known to offer a better sense of presence.

Visual cues capture the majority of our attention and are responsible for feeding maximum information to our brain. After vision, hearing is the second most important sense. Touch is usually less significant and rarely contributes to providing meaningful experiences. The senses of smell and taste, on account of them being very difficult to reproduce, are omitted from consideration in this case [19]. VR development majorly involves manipulating our senses via visual stimuli, followed by secondary contributions from sounds and noises.

Visual cues and the amount of interaction offered usually leads to an experience that is closer to reality. Ideally, we should be able to produce visuals that generate feedback that equals, if not exceeds, the human visual system [20]. The field of view (FOV) needs to be designed carefully, especially parts where the human is supposed to focus the most. The visual acuity, defined as the sharpness of viewing, should also be calibrated accordingly. The human eye can perceive a large spectrum of colours, ranging in 10 orders of magnitude [21]. None of the screens, as of now, can cover the whole spectrum. Appropriate colour-mapping techniques must be employed to achieve the best possible replication of the colour scheme [22].

There are other factors that contribute to the quality of a VR experience too [14]. Defects in the hardware, failing to provide perfect stimuli to the eyes, may generate a feeling of sickness. If they are not taken care of, they might cause simulator sickness [23]. However, there are other crucial design issues: system latency and frame rate variations. Higher frame rates are known to provide better stimuli to the eye while too much latency may cause the eyes to strain [24]. These issues are discussed in depth in later sections.

The production of audio stimulus is very different from traditional sound when it comes to VR. Given that immersion and interactivity are key characteristics of VR, special audio production techniques like spatial surround sound are used. In freer VR scenarios that provide six degrees of freedom, this audio needs to be programmed appropriately as well. For example, if the user is standing far from a barking dog, the audio should sound distant but if they are standing near the dog, the sound of the bark should seem closer. Such design factors contribute to the quality of the simulation a lot as well [25].

The level of realism, decided by the quality of the simulation, is very important in the use of VRET in order to accurately replace real-life scenarios with their virtual counterparts. The effectiveness of a particular VR simulation in aiding exposure therapy is directly proportional to the level of realism provided by it.

1.4 BENEFITS

VR has shown significant potential in exposure therapy in its comparison to traditional methods. Certain studies and meta-analyses have found evidence of greater or equivalent efficacy of VRET as compared to in vivo – the gold standard [26] for treatment of anxiety disorders [27–31]. Multiple studies even suggest that systematic delivery of VR with cognitive interventions, homework, questionnaires, surveys and other additional methods could dramatically increase the efficacy of VRET [32–34]. This points at the alteration and adaptation of delivery of therapy on the same principles when using VR for therapy.

1.4.1 LOWER REFUSAL AND DROP-OUT RATES

Often, confrontation of feared stimuli can be too intimidating or intense for a patient. This could lead to the patient either refusing treatment altogether or dropping out of treatment.

Previous research suggests that more patients would opt for VRET as compared to in-vivo exposure as a form of treatment, given the freedom to choose [35,36]. A phobia-specific study with a sample of 150 patients has shown that 27% refused in-vivo exposure, whereas only 3% refused VR-based exposure. Moreover, 76% chose VR-based exposure and 23.7% chose in-vivo exposure. Out of the set of patients that chose VR exposure, 90.4% claimed they chose it because they were afraid of confronting the feared object or situation in reality [37]. This shows a significantly stronger affinity of patients towards selecting VR-based exposure. Another study [38] suggests that people are more likely to stay in a treatment of their choosing.

A study with a sample of 352 U.S. soldiers suggests that VR may prove to address barriers to treatment. Majority of the soldiers, in the survey, mentioned they would be willing to use technology-based interventions for mental healthcare. It was also noted that 19% of those who said they would not be willing to talk to a mental health professional in person, agreed to access mental healthcare through VR [39].

Maples-Keller et al. [40] report that for certain disorders like autism spectrum disorder (ASD), studies have shown lower drop-out rates with the use of VR. Another meta-analysis found that for social anxiety, the dropout rate for virtual reality for anxiety disorders is 16%, slightly lower the 19.6% reported using another technique (cognitive-behavioural) [41] and the 19.7% reported by a meta-analysis of attrition from traditional therapy [42].

Overall, the first generation of studies provides empirical evidence to prove how VR could help lower refusal and attrition rates in therapy. Abandonment could lead to additional problems and this could help tremendously in helping a patient endure through the entire course of therapy. This advantage could also serve as an asset in clinical delivery of VRET in places where the general population is apprehensive about seeking help.

1.4.2 More control to Therapist

VRET affords complete control to the therapist who can expose the patient to the required stimuli in the doses they feel necessary. This is advantageous because it allows the therapist to tackle aspects of the simulation in a systematic manner, unlike in vivo. For example, for a person with entomophobia (the fear of insects), having to actually stand at a distance from insects while worrying about them flying, would cause an overwhelming reaction. VR eliminates this thought by limiting the possibilities of other anxiety-evoking triggers and focussing on the main one at hand. This serves as a safer way for patients to prepare for the feared stimulus in a real-world setting, without compromising on the level of realism [43]. VRET enables the therapists to see exactly what the environment is and what element within the virtual environment the patient is giving their attention to. Boeldt et al. [26] highlight that this addresses the following four limitations of imaginal exposure effectively:

1. not every patient imagines well;
2. the ability to form mental images declines with age;
3. the patient's imagery may be too frightening;
4. the therapist neither knows nor controls what is being imagined.

VR also broadens the scope to include modification and personalization in content. Therapists can provide a guided intervention directing the focus on certain aspects of the simulation that can potentially improve the efficacy of the treatment and increase the success of patient outcomes.

VR may also serve as an excellent tool to conduct further experimental research and controlled clinical trials in the fields of psychiatry and psychology because of the advantage of being able to control exposure stimuli and dosage [40]. Ressler et al. [44] examine the effects of combining D-cycloserine with exposure therapy on humans.

1.4.3 DATA COLLECTION AND ITS USES

Another potential area of the use of VR could be in the collection of data to improve the quality of treatment. With the use of VR head-mounted displayed and their additional hardware components, it is possible to collect relevant data for psychophysiological assessments, including specifications of exposure [40].

With the use of additional hardware or specialised VR head-mounted displays, we can also track eye movements. VR could potentially provide a major boost in the research and delivery of other therapy approaches like eye movement desensitization and reprocessing (EMDR); a psychotherapy approach designed to attenuate the severe affliction caused by traumatic memories [45]. It combines imaginal exposure with saccadic eye movements (rapid shifts of gaze that are normally executed with a single, smooth trajectory that ends near a selected visual target [46]) [47]. The client is made to focus on an external stimulus while they attend to brief doses of the trauma-related thoughts or material. They hold this thought in their memory and track the therapist's hand as it moves across various points in their field of vision [48]. This external stimulus often also includes hand-tapping or audio stimulation. With developments and documentations over time [49], EMDR has shown empirical results to effectively reduce physiological arousal, alleviate and reformulate negative beliefs and fast-track to results, amongst other advantages. The integration of VR in EMDR could enhance the external stimulus provided to the clients including auditory stimulation. The entire scenario or hints of the scenario could be re-created in the virtual environment and the guided eye movements would be provided within the scenario. Along with additional hardware, eye movement could also be tracked to provide data related to assessment to the therapist. However, proof of its efficacy has yet to be provided by research and empirical studies.

1.4.4 ADDITIONAL BENEFITS

Since the efficacy of VRET has proven to be equivalent, if not superior to that of the gold standard – in-vivo [29,30,50,50,50,51], it can serve as a widespread replacement for imaginal and in-vivo exposure. Therapists can leverage the fact that they do not have to leave the clinic or centre to deliver effective therapy on the spot without the need for prior planning in a case where the patient needs to be immersed in a higher- or lower-intensity simulation immediately. This leads to significantly reduced costs while increasing the feasibility of that stimulus. VRET also ensures high retention for the patient when combined with other psychological techniques. Another direct consequence of the fact that a superior immersive experience can be delivered within the four walls of a clinic can help ensure therapist-patient confidentiality [26,52,53]. Even when the patient is facing a stimulus that requires them to be in crowded areas, their identity is not revealed to outsiders. It is now possible to target specific aspects of the fear of flying example (refer Section 2) like take-off, landing and turbulence multiple times without special permissions, prior planning and additional costs, all while maintaining patient confidentiality.

1.5   VR IN THE USE OF SPECIFIC DISORDERS

Exposure therapy has been proven to be quite an effective treatment method for multiple psychiatric disorders, especially mild ones like fear and anxiety [54]. Emotional processing theory is the basis for contemporary exposure therapy. It suggests that fear memories are structures that contain information regarding fear stimuli, responses, and meaning. Hence, the goal of exposure therapy is to present concrete new information that satisfyingly disproves previously held beliefs, allow changes in emotions to be consequentially processed and ultimately alter the fear structures that form the foundation of the disorder. VRET provides the perfect environment for the patient to obtain new information related to their fears, where they can safely explore different ways of dealing with their fear stimuli and finding a comfortable way to deal with them.

1.5.1   PHOBIAS

The condition in which specific objects or situations (like flying, insects or needles) trigger fear or anxiety-inducing thoughts is characterized as specific phobia (SP). Usually, VRET of SPs involves simulating an environment in which the patient feels comfortable first. Slowly, as they proceed with their simulation, the objects or situations that trigger them are integrated. As they learn to deal with them, they become less and less anxious.

Many studies have investigated the efficiency of VRET on phobias in recent times. Most of them, conducted using RCTs with sufficient sample sizes, have found sizeable treatment effects and meaningful behavioural change. Moreover, data suggests that these effects are long-lasting [55].

The resulting information depicting effectiveness of VRET for specific phobias is very encouraging, especially for individuals fearing situations that may be unsafe, expensive, or unrealistic to conduct in vivo [40].

Moreover, it is known that less than 15–20% of those who suffer from SPs ever seek treatment. Apart from the effectiveness of the use of VR in treating phobias, a study provides empirical evidence to suggest that VR exposure therapy may prove valuable for increasing the number of phobias who seek treatment [35].

1.5.2   PANIC DISORDERS

Characterised by a sudden rush of anxiety manifested by physiological (e.g., heart palpitations, sweating, choking sensations) and cognitive (e.g., racing thoughts, fear of dying) symptoms, panic disorder and agoraphobia (PDA) lead to fear and/or avoidance of specific places or situations [56]. Usually, VRET for PDA would involve simulating scenes that trigger panic attacks, especially closed places like deserted highways, tunnels and parking lots.

While a lot of RCTs have praised VR-based cognitive behavioural therapy (CBT) for PDA, results vary between them relating to variations between VRET and traditional CBT. Some studies found that while VRET may deliver healthier treatment reaction [32,57] and a fewer number of sessions [33], long-term effects are very similar to those of traditional methods [14]. Long-term effects, particularly for 3 [32], 6 [58], 9 [59] and 12 months' [60] post-treatment have been found to be stable.

For panic disorder and agoraphobia, like specific phobias, VRET seems to perform better (at least early on in the treatment) if not equal to traditional CBT methods.

1.5.3 Anxiety Disorders

Generalized anxiety disorder (GAD) is a condition in which the patient experiences persistent, excessive and intrusive worrying to the extent that daily functioning becomes difficult [56]. The cause of worrying differs from person to person and could be anything, from stress about personal troubles to extreme worrying about climate change. Very few studies have investigated the use of VR simulations to inculcate relaxation techniques during treatment and they provide the initial foundation of feasibility for the usage of VRET in GAD cases [61]. One of the primary reasons of the smaller number of studies in GAD may be due to the fact that the cause of GAD is distinctive in every case and hence, it becomes very difficult to create a standardized VR simulation for the same. To battle this obstacle, VR simulations may focus on the most common worries between patients of GAD. One such example would be social anxiety disorder.

Social anxiety disorder (SAD) is a psychiatric condition in which the patient experiences anxiety in social interactions (e.g., conversations, meeting new people, public speaking) during which they might be judged or socially evaluated by others [56]. VRET simulations dealing with SAD usually build environments depicting social settings with virtual audiences like auditoriums and encourage the patient to voice their thoughts confidently in front of others.

Two RCTs involving such simulations found VRET to be equally helpful as traditional CBT [62] and better than control environments [63]. Studies researching stage fright and fear of public speaking have shown similar results [64–66], which were stable even 1 year post-treatment [64,67].

At the foundation of it, using VR simulations is more practical than making the patient interact with different kinds of people in different settings. Given that SAD affects 16 million adults in the United States alone [68], there is a lot of scope for VRET in helping with SAD.

Moreover, for all anxiety disorders, if it is not feasible to develop common VR simulations, they could be developed to encourage relaxation techniques, inculcate mindfulness and practice breathing exercises.

1.5.4 Post-Traumatic Stress Disorder

Post-traumatic stress disorder (PTSD) involves a history of exposure to a traumatic event as well as symptoms of intrusion, avoidance, negative alterations in cognitions and mood and alterations in arousal and reactivity [56]. VR simulations for PTSD usually involve simulating the setting of the traumatic event and slowly exposing the patient to a re-enactment of the traumatic event. Sometimes the trauma is very severe, and it may take several sessions and only after is the patient deemed fit enough, they are exposed to the re-enactment of trauma. The ultimate goal is to make the patient come to terms with what happened and making them deal with such an event in a virtual world is of a big help.

PTSD is one of the most researched areas of VRET. In fact, some of the earliest investigations into VRET have focused mainly on PTSD [69]. Early studies did not find much difference between VRET and traditional treatment methods but signs of improvement showed in patients that underwent VRET 6 months' post-treatment. While most of the research indicates that VRET is equally or more effective than traditional exposure therapy, one study found prolonged exposure therapy resulted in better results 3 and 6 months' post-treatment [70].

It is important to note that the result of VRET depends on a large range of factors, from content of the simulation to the nature of the therapist's involvement. Differences in VR environments, especially in their abilities to help in engaging emotions, contribute heavily to the result of the treatment. Overall, the vast amount of literature on VRET shows promise; however, it is worth mentioning that many studies had small sample sizes and/or lack comparison to other treatment techniques, among other things. Research with larger sample sizes along with appropriate comparisons shall help strengthen the foundation of VRET in the field of PTSD.

1.5.5   OBSESSIVE-COMPULSIVE DISORDERS

Obsessive-compulsive disorder (OCD) is a disorder in which people have recurring, unwanted thoughts, ideas or sensations (obsessions) that make them feel driven to do something repetitively (compulsions). The repetitive behaviours, such as hand washing, checking on things or cleaning, can significantly interfere with a person's daily activities and social interactions [56].

Not much research has been done in the field of OCD with respect to VR-based interventions. While there have been no RCTs conducted so far, there have been a couple of studies that investigated the acceptance of a virtual environment [71] [72]. These studies found that patients presented the same symptoms of OCD in the virtual environment as they would in reality.

Like GAD, the cause of OCD happens to be different for every patient, which makes creating a standardised simulation difficult. Also, VR may not be much needed for OCD, as the factors that trigger OCD in patients are usually easily found in reality or via imagination. However, VR may prove to be effective in cases where triggers are hard to find in real life and only further research can tell how much VR can help with that.

1.5.6   PAIN MANAGEMENT

Pain is an uncomfortable experience of emotions and senses that is felt most when one gives attention to it. Cognitive distraction is frequently used to manage pain, especially when undergoing painful medical procedures. VR has been found to be an excellent form of distraction, especially cases involving burn-related pain or physical therapy.

An investigation of four patients suffering from burns showed lower pain ratings when given occupational therapy via VR [73]. Another study comparing routine analgesia and analgesia coupled with a VR game found VR to be very helpful and more effective, as it distracts the patient from the feeling of numbness [74]. One such fMRI study that investigated brain activity relating to pain found that participants spent less time thinking about the pain along with decreased activity in regions corresponding to processing of pain-related emotions [75]. All such studies provide enough support to the fact that VR techniques can help in pain reduction.

VR can also be used to assist in dealing with chronic pain by learning and practicing pain management, coupled with traditional techniques. One such system was developed that made its users take a mindful stroll virtually as they learn meditative techniques that reduce stress and ease the sensation of pain. Initial results suggest that this system is more effective compared to achieving the same goals via traditional (control) techniques only [76].

However, specific factors that help VR achieve these levels of pain "reduction" are still unknown and continue to provide researchers an interesting field to investigate.

1.5.7 ADDICTION

When it comes to treatment of dependency and addiction, it is necessary to keep checking progress of the patient by exposing them to triggers (such as a bottle of alcohol, syringe, needle) and observing their reactions. Triggers in the form of real objects carry the risk of consumption and consequence relapse. VR can provide trigger objects that can be safely placed in virtual environments that do not pose this risk. A pilot study involving VRET via cues and triggers showed that participants were able to fight symptoms like bodily arousal [77]. A simulation hosting virtual "cocaine" was found to be similarly effective [78]. VR has been used to help patients de-addict from nicotine effectively too, by helping them control their cravings and not fall prey to various subjective cues. Moreover, the effect is not limited to just substance abuse and alcohol. A study involving a virtual casino showed that participants were able to dodge and control their symptoms like physiological arousal and the urge to gamble [79,80].

Many such studies involving treatment of addiction issues suggest that regular VR-based exposure therapy is effective in evoking reactions to cues and resulting craving, across various types of dependency problems.

1.5.8 MORE COMPLEX DISORDERS

The results of studies investigating mild disorders indicate that the technology has the potential to benefit more complex disorders too, like schizophrenia and autism.

Schizophrenia is a severe mental illness that includes psychotic symptoms (e.g., hallucinations, delusions), disruptions to normal emotional/behavioural functioning (e.g., flat affect, reduced pleasurable experiences, isolation), and difficulty with cognitive processing [56]. VR simulations would allow patients to practice their social skills and learn to cope in situations of social distress that are coupled with delusions. A small, randomized control trial discovered that patient showed decreased levels of delusions and better in-vivo social interactions after undergoing social skill training (SST) along with a VR system [81]. A few pilot studies have found similar results that support the efficacy of using VR to help combat schizophrenia [82,83]. A study found an increased interest in SST when it was coupled with VR, encouraging patients to undergo treatment [78].

Another complex disorder, autism spectrum disorder (ASD), is a developmental disorder marked by repetitive or restrictive patterns of behaviour and difficulties with social communication and interaction [40]. Research investigating the use of VR in the field of autism is very limited but does seem promising.

Preliminary results from a study involving autistic teenagers that used VR-based computer tasks to enhance social skills and interactions, have seen improved social performance [84]. However, it is worth noting that the study involved autistic teenagers whose intelligence was categorised as average or above, the sample size was very small (N = 8) and the system was designed in a special way to allow better communication with the challenged. A couple of studies also highlighted improvements with respect to recognition of emotion and enhancement of communication techniques along with other improvements [50].

VR needs to tread carefully in the waters of complex disorders as it is a relatively new and modern technology. However, given its promising results with mild disorders, a carefully designed VR simulation may go a long way in benefitting patients with such complicated conditions as well.

1.6   METHODS TO ADMINISTER VRET IN THERAPY

The use of VR in widespread clinical treatment is currently still in its nascent stage with experiments and proposals still defining a set standard and structure for delivery of VRET. An established framework could serve as an optimised method to reduce triggers, set protocols to handle triggers, induce relaxation, increase retention and reduce attrition rates. Several studies and experiments have addressed the same [32,33]. However, it is important to note that it may be very difficult, if not impossible, to generalise a single framework for all possible disorders that VR may be used to treat.

1.6.1   COMPONENTS OF VR CLINICAL SYSTEM

It is inadvisable to replace mental health professionals by a VRET system as they serve as the key component of this process. Their clinical skills serve as essential support mechanisms in delivering VRET and may be considered irreplaceable [85] The VR system component required to deliver VRET in clinics could potentially be as follows:

- The output tool that immerses the user in the virtual environment. Depending on the method of immersion, this could potentially be an HMD (phone or individual). This could include providing combinations of various stimulatory experiences including visual, auditory and haptic feedback.

- The input tools that allow the user to interact with the virtual environment while continually recording and reporting the movement of the patient. This would most commonly include hardware like remotes, data gloves, trackers or mice. For more sophisticated systems that could provide natural language processing capabilities to tune the elements of a virtual environment based on verbal cues, an internal or external microphone would be used.
- The therapist's interface (software) that serves as the control panel for therapeutic habituation by enabling anxiety modulation capabilities using VR. The importance of the clinician's interface has been highlighted by Rizzo et al [86]. This ability to control the required triggers involved in the simulation in real time provides great flexibility, functionality and customisation potential to improve the quality of VRET.

A major advantage is that VR is compatible with multiple kinds of clinical approaches to treat a variety of mental health disorders: cognitive, experiential or behavioural [87]. There are controlled trials and studies that are critical in shaping research in this field. They majorly focus on two psychological approaches: experiential cognitive approach [32,33,88–92] and cognitive behavioural approach [4,36,52,93–96]. These approaches, while having disparities, broadly refer to providing the patient with a graded exposure dosage in VR.

Broadly, the treatment course for therapy using VR would include beginning with understanding the link between the patient and the disorder. This could include a background check involving psychoeducation about the disorder and the patient history (possible trauma, triggers) [40]. The therapist gauges the avoidance strategies, best described as a temporary solution to a long-term problem. It is a coping mechanism that may be used by patients in a bid to escape an uncomfortable feeling, thought or scenario. This pushes them away from addressing the actual issue at hand.

Vincelli et al. [32] approach the session by subjecting the patients to scenarios that may be relevant to the disorder and then having the patient report their experience on the SUDs scale. This can help create the hierarchy of virtual environments. The next step is to establish a hierarchy of stimuli within the selected virtual environment. Once these parameters are set, the therapist can begin session-wise treatment. Between-session interventions (also known as homework) are important especially for VR in therapy. Benbow and Anderson provide empirical evidence hinting at their potential to reduce attrition rates [97].

Each subsequent session can begin by checking on homework followed by gradation of experience. This can help set the difficulty level for the simulation for that session and gauge the progress. Based on the approach, the therapist can then move on to cognitive restructuring through VR. Depending on the quality and availability of a customisable VR system, each exposure session can be individualised for each patient. This is where the therapist's interface comes in. As the simulation progresses, the therapist can tune the parameters of the exposure as well as control the location, nature, intensity, timing and duration of the stimulus to be provided. For example, in the example of stage fright (Table 1.1, Section 2), the therapists can control the number of people engaging in conversation, the intensity of applause or other such potential triggers based on the level of tolerance of the patient. Special emphasis can be put on background auditory elements like indistinct surrounding conversations, breathing sounds, cars passing by, doors opening and closing, etc. The patient can be put through the virtual environment repeatedly until they report lower distress rates based on the therapist's observations and the SUDs scale. The patient can move to a higher difficulty level when both the patient and the therapist feel that they are ready. Therapists can even look at the effects of coupling VR-based therapy with imaginal and feasible in-vivo components. Depending on the judgement of the therapist, the patient can be called in for follow-up sessions in the future as they help with retention. However, there is a need for training therapists in the use of VR in clinical treatment due to the deviations from traditional therapeutic protocols.

However, this is not the only proposed approach or even the best approach. Research suggests that trials and studies have not become methodologically rigorous [50,84,98,99] over the years as they should have. Additional research is required to affirm the role of VR in clinical treatments.

1.7   ISSUES AND RESOLUTION

VRs use in therapy does not come without its technical and therapeutic hindrances. This section explores the possible technical and physical limitations, their solutions (if any).

One of the primary concerns with any VR-based application is cybersickness, which is an unwanted side effect of using immersive interfaces. It is triggered by moving visual stimuli and has symptoms similar to motion sickness [100]. It causes a broad range of discomforting reactions the likes of nausea, dizziness, tiredness, fatigue, vomiting, double vision, decreased hearing, headaches, seizures and disorientation [101]. There may be no permanent fix to this problem, but it can be reduced with regulated usage of VR during therapy. This is not just based on duration of immersion, but also on other factors like refresh rates, method of delivery, comfort of HMD, etc. The therapist can use predesigned questionnaires like visually induced motion sickness susceptibility questionnaire (VIMSSQ) to gauge how susceptible a potential user may be to cybersickness. VIMSSQ specifically is based on the past use of visual devices by the respondent and any incidences of dizziness, eyestrain, nausea, fatigue and headache they may have had [102]. While studies point towards decreased rates of cybersickness with newer improved hardware [103], the disparities are not drastic enough to be able to rule out that this issue may be a hindrance in the adoption of VR as a therapeutic tool.

Similar to video games, VR provides a platform for its users to play around in a fictional environment. Hence, it poses another area of concern – the possibility of addiction and social isolation, especially amongst younger generations [104]. In this environment, there is also the possibility of a patient experiencing a loss of reality. This may encourage more dangerous behaviour if the patient is unable to distinguish the real world and a virtual environment, where they are safe under the assumption that their actions will not have any real-life consequences. Such a danger is more prominent amongst people with pathological disorders like personality disorders or schizophrenia, who have a tendency to get disoriented by switching between real and virtual worlds [28].

Quality VRET is personalised for the patient. In order to boost efficacy rates and provide surety of treatment, the content library must be vast. In spite of introduction of newer technology and consistent reduction of headset costs, development of accurate virtual environments is computationally heavy, requires a certain skillset for creation and is significantly expensive. This may pose as an obstacle in adoption of VRET as a preferred form of treatment.

When in use, technological glitches may pose as an impediment and trouble-shooting them is essential for a smooth therapy session. Therapists may also not be used to delivering effective therapy which is why initial training may be necessary in order to alleviate any technology-related problems.

Visual and auditory stimulation increase the chances for a patient to be triggered by an element of any simulation. There is a need to pay special attention to this because a trigger treated wrongly has the potential to become counterproductive for the therapy session. Traditional exposure therapy provides systems for patient assessment before they are subjected to any fear-inducing situations. This alone may not provide a therapist complete confidence over a new system like VR. VR-based systems coupled with biofeedback sensor data provide a great way for the therapist to gauge the physicality and emotional changes in the client. This could include eye tracking to measure pupil dilation, galvanic skin response (GSR) for skin conductance, electro-encephalogram (EEG) and functional magnetic resonance imaging (fMRI) for brain activity, electrocardiogram (ECG) for heart rate and facial expressions.

Accessibility has always been a concern for technology like virtual reality. Manufacturing of headsets and development of simulations via computer graphics and animation is expensive. However, with recent advancements in filming, this cost can go down. Three hundred sixty–degree cameras can be employed to shoot real-life scenarios and these films can be played in a VR headset to administer VRET. Given that characters and objects in animated simulations are created using computer graphics, such simulations may offer lesser levels of realism. While 360-degree videos may be able to overcome this obstacle by providing lifelike scenarios as captured by cameras, they offer movement only in three degrees of freedom and, hence, may limit interactivity with objects. It is up to the designer and therapist to decide the form of simulation based on the requirements of the treatment and the patient. A study conducted by Trine et al [105] involved three such 360-degree videos filmed at a shopping centre that aimed to help users get over their social anxiety. The participants of the study reported increased levels of anxiety and presence throughout their treatment course and recommended that VRET be in-tegrated before in-vivo exposure.

There may also be some deep-rooted issues with gamification, for environments that may include a reward system on completion of certain tasks. Similar to video game addiction [106], the provided reinforcements may lead to an unsolicited dependency and an unwanted fixation to the environment. In cases where gamification can enhance therapy, an arbitrary therapist-devised reward system may foster non-addictive positive reinforcement as encouragement to patients to overcome tougher tasks.

1.8 DISCUSSION

Apart from the technical benefits VRET can provide, it is also interesting to dive into the societal impact it can have. VR has shown potential in reducing the time taken for treatment. Based on the empirical evidence that further research may provide us in solidifying the basis of this statement, VRET could help patients as well as mental health professionals:

- Reduction of cost: Mental health awareness is especially low amongst people lower on the socio-economic scale. Seeking mental health treatment is known to be costly and is charged per session. Reduction in time of treatment is directly proportional to the cost the patient bears. This could potentially encourage more people to seek help.
- Accommodation of additional patients: In the time saved by VRET, therapists can accommodate additional patients. Hence, shorter time for treatment does not mean loss of revenue for psychologists.

Additionally, if future studies can provide more evidence of a lower attrition rate being associated with VR, widespread adoption of VR may help reduce relapse in patients.

Most of the research in the field of VRET has been concentrated in developed countries like the United States and the United Kingdom. The adaptations, beliefs and challenges associated with psychotherapy vary from culture to culture [107]. Further research needs to be conducted in populations of different cultures, especially in non-Western countries like India, as the population's psychologies differ a lot [108]. A survey conducted by the authors found that most psychologists are excited about the introduction of VRET and are eager to test the technology out. Consequently, further research is being conducted.

It is estimated that 1 in every 10 people may need mental healthcare at some point in their life [109]. Knowing that, it is terrifying to learn that the ratio of mental health professionals to the general population can be as low as 2:100,000. Of the 65 countries that have established policies and plans related to mental health, they have often failed to enforce them, especially in spreading awareness about the most common conditions and reducing the stigma around it. In countries where the mental health infrastructure is so fragile, the introduction of VRET will not only reduce the pressure on mental health professionals by reducing cost, treatment time, and chances of drop-out and relapse but will also attract the general population to take an interest in the field.

It is important to note that the technology is simply not a direct answer to the need of an improvement in exposure therapy. The content delivered by the technology plays a major role in the outcome of the therapy [72,110]. While the authors advocate for the use of this technology, there is a need for further research as studies have usually been small, causing negative results less likely to be reported. Compared to the potential of it, the technology has not been applied enough to mental health.

VR can be used not just for treatment, but for assessment of psychiatric symptoms as well. Though the technology has the potential of becoming the benchmark in assessment, very few reliable and robust tests have been conducted related to it [43]. Even as a form of treatment, VR can be used to innovate so many treatment methods. Many common disorders, like depression, remain relatively less explored with respect to VR-based treatment. While exposure therapy remains a common and simple form of treatment, the innovative approach of VR can be applied to many more treatment techniques, especially complex ones that are used for disorders like schizophrenia and autism.

With leaping innovations in artificial intelligence, the possibility of a fully automated VR-based therapy, that might eliminate the need of a therapist's presence, is still intriguing. Technologies like augmented reality can also assist in the treatment, and add several more features and benefits to treatments.

Our review offers a look into the current scenario and a small glimpse of the future of mental health treatments. VR for mental health is still in its early days. Simulations are quite limited in quantity. They lack features, special training is needed for operation, if not creation of, suitable environments, and simulator sickness still remains a hurdle that needs to be overcome carefully with content and hardware design. However, the technology is developing fast and these are probably short-term concerns.

As VR is able to simulate scenarios that are not easily found in the real world, maybe it will help deliver treatment results that are not easily found with traditional techniques as well. VR has revolutionised many fields, and it is time mental health got a taste of it too.

REFERENCES

[1] M. C. Hsieh and J. J. Lee, "Preliminary Study of VR and AR Applications in Medical and Healthcare Education," *Journal of Nursing and Health Studies*, vol. 03, no. 1, p. 1, 2018.

[2] L. Gregg and N. Tarrier, "Virtual Reality in Mental Health," *Social Psychiatry and Psychiatric Epidemiology*, vol. 42, pp. 343–354, 3, 2007.

[3] R. T. da Costa, M. R. de Carvalho and A. E. Nardi, "Virtual Reality Exposure Therapy in the Treatment of Driving Phobia," *Psicologia: Teoria e Pesquisa*, vol. 26, pp. 131–137, 3, 2010.

[4] B. O. Rothbaum, L. F. Hodges, R. Kooper, D. Opdyke, J. S. Williford and M. North, "Effectiveness of Computer-Generated (virtual reality) Graded Exposure in the Treatment of Acrophobia," *American Journal of Psychiatry*, vol. 152, pp. 626–628, 1995.

[5] Fortune Business Insights, "Virtual Reality (VR) in Healthcare Market Size, Share & Industry Analysis," 2019.

[6] S. A. Steinman, B. M. Wootton and D. F. Tolin, "Exposure Therapy for Anxiety Disorders," in H. S. Friedman (Ed.), *Encyclopedia of Mental Health*, Elsevier, 2016, pp. 186–191.

[7] S. Taylor, "Exposure," in M. Hersen and W. Sledge (Eds.), *Encyclopedia of Psychotherapy*, Elsevier, 2002, pp. 755–759.

[8] J. Wolpe, *The Practice of Behavior Therapy*, 4th ed., Elmsford, NY: Pergamon Press, 1990, pp. xvi, 421–xvi, 421.

[9]   S. L. Johnson, "Transtheoretical and Multimodal Interventions," in S. L. Johnson
      (Ed.), Therapist\textquotesingles Guide to Posttraumatic Stress Disorder Intervention,
      Elsevier, 2009, pp. 123–169.
[10] K. B. Wolitzky-Taylor, J. D. Horowitz, M. B. Powers and M. J. Telch,
      "Psychological Approaches in the Treatment of Specific Phobias: A Meta-analysis,"
      Clinical Psychology Review, vol. 28, pp. 1021–1037, 7, 2008.
[11]   M. A. Tompkins, "Nuts and Bolts of Imaginal Exposure," [Online]. Available:
      http://sfbacct.com/from-ocd-to-anxiety/nuts-and-bolts-of-imaginal-exposure/.
[12] C. Miller, "Flooding," in M. Herson and W. Sledge (Eds.), Encyclopedia of
      Psychotherapy, Elsevier, 2002, pp. 809–813.
[13] C. H. Rankin, T. Abrams, R. J. Barry, S. Bhatnagar, D. F. Clayton, J. Colombo, G.
      Coppola, M. A. Geyer, D. L. Glanzman, S. Marsland, F. K. McSweeney, D. A.
      Wilson, C.-F. Wu and R. F. Thompson, "Habituation Revisited: An Updated and
      Revised Description of the Behavioral Characteristics of Habituation,"
      Neurobiology of Learning and Memory, vol. 92, pp. 135–138, 2009.
[14] A. Pelissolo, M. Zaoui, G. Aguayo, N. Sai, Yao, S. Roche, R. Ecochard, C. Pull, A.
      Berthoz, R. Jouvent and J. Cottraux, "Virtual Reality Exposure Therapy Versus
      Cognitive Behavior Therapy for Panic Disorder With Agoraphobia: A Randomized
      Comparison Study," Journal of Cybertherapy and Rehabilitation, vol. 5, no. 1,
      pp. 35, 2012.
[15] P. Lindner, A. Rozental, A. Jurell, L. Reuterskiöld, G. Andersson, W. Hamilton, A.
      Miloff and P. Carlbring, "Experiences of Gamified and Automated Virtual Reality
      Exposure Therapy for Spider Phobia: Qualitative Study," JMIR Serious Games, vol.
      8, p. e17807, 4, 2020.
[16] R. Holloway and A. Lastra, "Virtual Environments: A Survey of the Technology,"
      1993. Tech. Report TR93–033. University of North Carolina, Chapel Hill.
[17] M. Slater and S. Wilbur, "A Framework for Immersive Virtual Environments
      (FIVE): Speculations on the Role of Presence in Virtual Environments," Presence:
      Teleoperators and Virtual Environments, vol. 6, pp. 603–616, 12, 1997.
[18] H. L. Miller and N. L. Bugnariu, "Level of Immersion in Virtual Environments
      Impacts the Ability to Assess and Teach Social Skills in Autism Spectrum Disorder,"
      Cyberpsychology, Behavior, and Social Networking, vol. 19, pp. 246–256, 4, 2016.
[19] M. L. Heilig, "EL Cine del Futuro: The Cinema of the Future," Presence:
      Teleoperators & Virtual Environments, vol. 1, pp. 279–294, 1992.
[20] J. Helman, "Performance Requirements and Human Factors," SIGGRAPH. 1995.
[21] S. Hüttermann, N. J. Smeeton, P. R. Ford and A. M. Williams, "Color Perception
      and Attentional Load in Dynamic, Time-Constrained Environments," Frontiers in
      Psychology, vol. 9, 1, 2019. https://doi.org/10.3389/fpsyg.2018.02614
[22] H. S. Faridul, T. Pouli, C. Chamaret, J. Stauder, A. Tremeau and E. Reinhard, A
      Survey of Color Mapping and Its Applications, The Eurographics Association, 2014.
[23] J. Lee, M. Kim and J. Kim, "A Study on Immersion and VR Sickness in Walking
      Interaction for Immersive Virtual Reality Applications," Symmetry, vol. 9, p. 78, 2017.
[24] L. Sidenmark, N. Kiefer and H. Gellersen, "Subtitles in Interactive Virtual Reality:
      Using Gaze to Address Depth Conflicts," in Workshop on Emerging Novel Input
      Devices and Interaction Techniques, 2019.
[25] Z. Yan, J. Wang and Z. Li, "A Multi-criteria Subjective Evaluation Method for
      Binaural Audio Rendering Techniques in Virtual Reality Applications," in 2019
      IEEE International Conference on Multimedia & Expo Workshops (ICMEW), 2019.
[26] D. Boeldt, E. McMahon, M. McFaul and W. Greenleaf, "Using Virtual Reality
      Exposure Therapy to Enhance Treatment of Anxiety Disorders: Identifying Areas of
      Clinical Adoption and Potential Obstacles," Frontiers in Psychiatry, vol. 10, 2019.
      https://doi.org/10.3389/fpsyt.2019.00773

[27] B. K. Wiederhold, D. P. Jang, R. G. Gevirtz, S. I. Kim, I. Y. Kim and M. D. Wiederhold, "The Treatment of Fear of Flying: A Controlled Study of Imaginal and Virtual Reality Graded Exposure Therapy," *IEEE Transactions on Information Technology in Biomedicine*, vol. 6, pp. 218–223, 9, 2002.

[28] A. Gorini and G. Riva, "Virtual Reality in Anxiety Disorders: The Past and the Future," *Expert Review of Neurotherapeutics*, vol. 8, pp. 215–233, 2, 2008.

[29] M. B. Powers and P. M. G. Emmelkamp, "Virtual Reality Exposure Therapy for Anxiety Disorders: A Meta-analysis," *Journal of Anxiety Disorders*, vol. 22, pp. 561–569, 4, 2008.

[30] T. D. Parsons and A. A. Rizzo, "Affective Outcomes of Virtual Reality Exposure Therapy for Anxiety and Specific Phobias: A Meta-analysis," *Journal of Behavior Therapy and Experimental Psychiatry*, vol. 39, pp. 250–261, 9, 2008.

[31] C. Suso-Ribera, J. Fernández-Álvarez, A. García-Palacios, H. G. Hoffman, J. Bretón-López, R. M. Baños, S. Quero and C. Botella, "Virtual Reality, Augmented Reality, and In Vivo Exposure Therapy: A Preliminary Comparison of Treatment Efficacy in Small Animal Phobia," *Cyberpsychology, Behavior, and Social Networking*, vol. 22, p. 31–38, 1, 2019.

[32] F. Vincelli, L. Anolli, S. Bouchard, B. K. Wiederhold, V. Zurloni and G. Riva, "Experiential Cognitive Therapy in the Treatment of Panic Disorders with Agoraphobia: A Controlled Study," *CyberPsychology & Behavior*, vol. 6, pp. 321–328, 6, 2003.

[33] Y.-H. Choi, F. Vincelli, G. Riva, B. K. Wiederhold, J.-H. Lee and K.-H. Park, "Effects of Group Experiential Cognitive Therapy for the Treatment of Panic Disorder with Agoraphobia," *CyberPsychology & Behavior*, vol. 8, pp. 387–393, 8, 2005.

[34] T. F. Wechsler, F. Kümpers and A. Mühlberger, "Inferiority or Even Superiority of Virtual Reality Exposure Therapy in Phobias?—A Systematic Review and Quantitative Meta-Analysis on Randomized Controlled Trials Specifically Comparing the Efficacy of Virtual Reality Exposure to Gold Standard in vivo Exposure in Agoraphobia, Specific Phobia, and Social Phobia," *Frontiers in Psychology*, vol. 10, 9, 2019.

[35] A. Garcia-Palacios, H. G. Hoffman, S. K. See, A. Tsai and C. Botella, "Redefining Therapeutic Success with Virtual Reality Exposure Therapy," *CyberPsychology & Behavior*, vol. 4, pp. 341–348, 6, 2001.

[36] A. Garcia-Palacios, H. Hoffman, A. Carlin, T. A. Furness and C. Botella, "Virtual Reality in the Treatment of Spider Phobia: A Controlled Study," *Behaviour Research and Therapy*, vol. 40, pp. 983–993, 9, 2002.

[37] A. Garcia-Palacios, C. Botella, H. Hoffman and S. Fabregat, "Comparing Acceptance and Refusal Rates of Virtual Reality Exposure vs. In Vivo Exposure by Patients with Specific Phobias," *CyberPsychology & Behavior*, vol. 10, pp. 722–724, 2007.

[38] D. Steidtmann, R. Manber, B. A. Arnow, D. N. Klein, J. C. Markowitz, B. O. Rothbaum, M. E. Thase and J. H. Kocsis, "Patient Treatment Preference as a Predictor of Response and Attrition in Treatment for Chronic Depression," *Depression and Anxiety*, vol. 29, pp. 896–905, 2012.

[39] J. A. B. Wilson, K. Onorati, M. Mishkind, M. A. Reger and G. A. Gahm, "Soldier Attitudes about Technology-Based Approaches to Mental Health Care," *CyberPsychology & Behavior*, vol. 11, pp. 767–769, 2008.

[40] J. L. Maples-Keller, B. E. Bunnell, S.-J. Kim and B. O. Rothbaum, "The Use of Virtual Reality Technology in the Treatment of Anxiety and Other Psychiatric Disorders," *Harvard Review of Psychiatry*, vol. 25, pp. 103–113, 2017.

[41] E. Fernandez, D. Salem, J. K. Swift and N. Ramtahal, "Meta-Analysis of Dropout From Cognitive Behavioral Therapy: Magnitude, Timing, and Moderators," *Journal of Consulting and Clinical Psychology*, vol. 83, pp. 1108–1122, 2015.

[42] J. K. Swift and R. P. Greenberg, "Premature Discontinuation in Adult Psychotherapy: A Meta-analysis," *Journal of Consulting and Clinical Psychology*, vol. 80, pp. 547–559, 2012.

[43] D. Freeman, S. Reeve, A. Robinson, A. Ehlers, D. Clark, B. Spanlang and M. Slater, "Virtual Reality in the Assessment, Understanding, and Treatment of Mental Health Disorders," *Psychological Medicine*, vol. 47, pp. 2393–2400, 3 2017.

[44] K. J. Ressler, B. O. Rothbaum, L. Tannenbaum, P. Anderson, K. Graap, E. Zimand, L. Hodges and M. Davis, "Cognitive Enhancers as Adjuncts to Psychotherapy," *Archives of General Psychiatry*, vol. 61, pp. 1136, 11, 2004.

[45] F. Shapiro, "Efficacy of the Eye Movement Desensitization Procedure in the Treatment of Traumatic Memories," *Journal of Traumatic Stress*, vol. 2, pp. 199–223, 4, 1989.

[46] E. L. Keller, B.-T. Lee and K.-M. Lee, "Frontal Eye Field Signals That May Trigger the Brainstem Saccade Generator," in Keller E. L., Lee B.-T. and Lee K.-M. (Eds.), *Progress in Brain Research*, Elsevier, 2008, pp. 107–114.

[47] L. H. Jaycox and E. B. Foa, "Post-traumatic Stress Disorder," in A.S. Belleck and M. Hersen (Eds.), *Comprehensive Clinical Psychology*, Elsevier, 1998, pp. 499–517.

[48] "What Is EMDR?," [Online]. Available: https://www.emdr.com/what-is-emdr/.

[49] F. Shapiro, *Eye Movement Desensitization and Reprocessing: Basic Principles, Protocols, and Procedures*, 2nd ed., New York, NY, US: Guilford Press, 2001, pp. xxiv, 472–xxiv, 472.

[50] K. Meyerbröker and P. M. G. Emmelkamp, "Virtual Reality Exposure Therapy in Anxiety Disorders: A Systematic Review of Process-and-outcome Studies," *Depression and Anxiety*, vol. 27, pp. 933–944, 8, 2010.

[51] D. Opriş, S. Pintea, A. García-Palacios, C. Botella, Ş. Szamosközi and D. David, "Virtual Reality Exposure Therapy in Anxiety Disorders: A Quantitative Meta-analysis," *Depression and Anxiety*, vol. 29, pp. 85–93, 11, 2011.

[52] B. O. Rothbaum, P. Anderson, E. Zimand, L. Hodges, D. Lang and J. Wilson, "Virtual Reality Exposure Therapy and Standard (in Vivo) Exposure Therapy in the Treatment of Fear of Flying," *Behavior Therapy*, vol. 37, pp. 80–90, 3, 2006.

[53] A. Pittig, R. Kotter and J. Hoyer, "The Struggle of Behavioral Therapists With Exposure: Self-Reported Practicability, Negative Beliefs, and Therapist Distress About Exposure-Based Interventions," *Behavior Therapy*, vol. 50, pp. 353–366, 3, 2019.

[54] B. Bandelow, M. Reitt, C. Röver, S. Michaelis, Y. Görlich and D. Wedekind, "Efficacy of Treatments for Anxiety Disorders," *International Clinical Psychopharmacology*, vol. 30, pp. 183–192, 7, 2015.

[55] C. Thng, N. Lim-Ashworth, B. Poh and C. G. Lim, "Recent Developments in The Intervention of Specific Phobia Among Adults: A Rapid Review," *F1000Research*, vol. 9, p. 195, 3, 2020.

[56] A. P. Association, *Diagnostic and Statistical Manual of Mental Disorders (DSM-5 (R))*, American Psychiatric Association Publishing, 2013.

[57] E. Malbos, R. M. Rapee and M. Kavakli, "A Controlled Study of Agoraphobia and the Independent Effect of Virtual Reality Exposure Therapy," *The Australian and New Zealand journal of psychiatry*, vol. 47, no. 2, pp. 160–168, 2 2013.

[58] C. Botella, A. García-Palacios, H. Villa, R. M. Baños, S. Quero, M. Alcañiz and G. Riva, "Virtual Reality Exposure in the Treatment of Panic Disorder and Agoraphobia: A Controlled Study," *Clinical Psychology & Psychotherapy*, vol. 14, pp. 164–175, 2007.

[59] A. Gorini, F. Pallavicini, D. Algeri, C. Repetto, A. Gaggioli and G. Riva, "Virtual Reality in the Treatment of Generalized Anxiety Disorders," *Studies in health technology and informatics*, vol. 154, pp. 39–43, 2010.

[60] A. Belloch, E. Cabedo, C. Carrió, J. A. Lozano-Quilis, J. A. Gil-Gómez and H. Gil-Gómez, "Virtual Reality Exposure for OCD: Is It Feasible? [Exposición mediante realidad virtual para el TOC: ¿Es factible?]," *Revista de Psicopatología y Psicología Clínica*, vol. 19, p. 37, 9, 2014.

[61] K. Kim, C.-H. Kim, K. R. Cha, J. Park, K. Han, Y. K. Kim, J.-J. Kim, I. Y. Kim and S. I. Kim, "Anxiety Provocation and Measurement Using Virtual Reality in Patients with Obsessive-Compulsive Disorder," *CyberPsychology & Behavior*, vol. 11, pp. 637–641, 12, 2008.

[62] G. Robillard, S. Bouchard, S. Dumoulin, T. Guitard and E. Klinger, "Using Virtual Humans to Alleviate Social Anxiety: Preliminary Report from a Comparative Outcome Study," *Studies in health technology and informatics*, vol. 154, pp. 57–60, 2010.

[63] P. L. Anderson, M. Price, S. M. Edwards, M. A. Obasaju, S. K. Schmertz, E. Zimand and M. R. Calamaras, "Virtual Reality Exposure Therapy for Social Anxiety Disorder: A Randomized Controlled Trial," *Journal of consulting and clinical psychology*, vol. 81, no. 5, pp. 751–760, 10, 2013.

[64] S. R. Harris, R. L. Kemmerling and M. M. North, "Brief Virtual Reality Therapy for Public Speaking Anxiety," *Cyberpsychology & Behavior: The Impact of the Internet, Multimedia and Virtual Reality on Behavior and Society*, vol. 5, no. 6, pp. 543–550, 12, 2002.

[65] H. S. Wallach, M. P. Safir and M. Bar-Zvi, "Virtual Reality Cognitive Behavior Therapy for Public Speaking Anxiety: A Randomized Clinical Trial," *Behavior Modification*, vol. 33, no. 3, pp. 314–338, 5, 2009.

[66] M. P. Safir, H. S. Wallach and M. Bar-Zvi, "Virtual Reality Cognitive-behavior Therapy for Public Speaking Anxiety: One-year Follow-up," *Behavior Modification*, vol. 36, no. 2, pp. 235–246, 3, 2012.

[67] I. Alsina-Jurnet, C. Carvallo-Beciu and J. Gutiérez-Maldonado, "Validity of Virtual Reality as a Method of Exposure in the Treatment of Test Anxiety," *Behavior Research Methods*, vol. 39, pp. 844–851, 12, 2007.

[68] Anxiety and Depression Association of America, "Facts & Statistics," [Online]. Available: https://adaa.org/about-adaa/press-room/facts-statistics. [Accessed 9 11 2020].

[69] B. O. Rothbaum, L. F. Hodges, D. Ready, K. Graap and R. D. Alarcon, "Virtual Reality Exposure Therapy for Vietnam Veterans With Posttraumatic Stress Disorder," *The Journal of Clinical Psychiatry*, vol. 62, pp. 617–622, 8 2001.

[70] C. Pitti, J. Bethencourt-Pére, J. Fuente, G. Ramón and W. Peñate, "The Effects of a Treatment Based on the Use of Virtual Reality Exposure and Cognitive-behavioral Therapy Applied to Patients with Agoraphobia," *International Journal of Clinical and Health Psychology*, vol. 8, no. 1, pp. 5–22, 2008.

[71] J. Ku, K. Han, H. R. Lee, H. J. Jang, K. U. Kim, S. H. Park, J. J. Kim, C. H. Kim, I. Y. Kim and S. I. Kim, "VR-Based Conversation Training Program for Patients With Schizophrenia: A Preliminary Clinical Trial," *Cyberpsychology & Behavior: The Impact of the Internet, Multimedia and Virtual Reality on Behavior and Society*, vol. 10, no. 4, pp. 567–574, 8, 2007.

[72] D. Freeman, J. Bradley, A. Antley, E. Bourke, N. DeWeever, N. Evans, E. Černis, B. Sheaves, F. Waite, G. Dunn, M. Slater and D. M. Clark, "Virtual Reality in the Treatment of Persecutory Delusions: Randomised Controlled Experimental Study Testing How to Reduce Delusional Conviction," *British Journal of Psychiatry*, vol. 209, no. 1, pp. 62–67, 7, 2016.

[73] D. A. Das, K. A. Grimmer, A. L. Sparnon, S. E. Mcrae and B. H. Thomas, "The Efficacy of Playing a Virtual Reality Game in Modulating Pain for Children With Acute Burn Injuries: A Randomized Controlled Trial [ISRCTN87413556]," *BMC Pediatrics*, vol. 5, no. 1, pp. 1, 3, 2005.

[74]   F. J. Keefe, D. A. Huling, M. J. Coggins and D. F. Keefe, "Virtual Reality for Persistent Pain: A New Direction for Behavioral Pain Management," *Pain*, vol. 153, pp. 2163–2166, 7, 2012.

[75]   H. G. Hoffman, T. L. Richards, B. Coda, A. R. Bills, D. Blough, A. L. Richards and S. R. Sharar, "Modulation of Thermal Pain-related Brain Activity with Virtual Reality: Evidence from fMRI," *Neuroreport*, vol. 15, no. 8, pp. 1245–1248, 6, 2004.

[76]   D. Gromala, X. Tong, A. Choo, M. Karamnejad and C. D. Shaw, "The Virtual Meditative Walk," in *Proceedings of the 33rd Annual ACM Conference on Human Factors in Computing Systems - CHI \textquotesingle15*, 2015.

[77]   J.-S. Choi, S. Park, J.-Y. Lee, H.-Y. Jung, H. W. Lee, C.-H. Jin and D.-H. Kang, "The Effect of Repeated Virtual Nicotine Cue Exposure Therapy on the Psychophysiological Responses: A Preliminary Study," *Psychiatry Investigation*, vol. 8, no. 2, pp. 155–160, 6, 2011.

[78]   D. G. Y. Thompson-Lake, K. N. Cooper, J. J. Mahoney, P. S. Bordnick, R. Salas, T. R. Kosten, J. A. Dani and R. De La Garza, "Withdrawal Symptoms and Nicotine Dependence Severity Predict Virtual Reality Craving in Cigarette-Deprived Smokers," *Nicotine & Tobacco Research: Official Journal of the Society for Research on Nicotine and Tobacco*, vol. 17, no. 7, pp. 796–802, 7, 2015.

[79]   J. H. Yoon, T. F. Newton, C. N. Haile, P. S. Bordnick, R. E. Fintzy, C. Culbertson, J. J. Mahoney, R. Y. Hawkins, K. R. LaBounty, E. L. Ross, A. I. Aziziyeh and R. D. La Garza, "Effects of D-cycloserine on Cue-induced Craving and Cigarette Smoking Among Concurrent Cocaine- and Nicotine-dependent Volunteers," *Addictive Behaviors*, vol. 38, no. 2, pp. 1518–1526, 2, 2013.

[80]   C. Perpiñá, C. Botella, R. Baños, H. Marco, M. Alcañiz and S. Quero, "Body Image and Virtual Reality in Eating Disorders: Is Exposure to Virtual Reality More Effective than the Classical Body Image Treatment?," *CyberPsychology & Behavior*, vol. 2, pp. 149–155, 4, 1999.

[81]   M. Rus-Calafell, J. Gutiérrez-Maldonado and J. Ribas-Sabaté, "A Virtual Reality-integrated Program for Improving Social Skills in Patients With Schizophrenia: A Pilot Study," *Journal of Behavior Therapy and Experimental Psychiatry*, vol. 45, no. 1, pp. 81–89, 3, 2014.

[82]   K.-m. Park, J. Ku, S.-h. Choi, H.-j. Jang, J.-y. Park, S. I. Kim And J.-j. Kim, "a Virtual Reality Application in Role-plays of Social Skills Training for Schizophrenia: A Randomized, Controlled Trial," *Psychiatry Research*, vol. 189, no. 2, pp. 166–172, 9, 2011.

[83]   H. G. Hoffman, D. R. Patterson, G. J. Carrougher and S. R. Sharar, "Effectiveness of Virtual Reality-based Pain Control With Multiple Treatments," *The Clinical Journal of Pain*, vol. 17, no. 3, pp. 229–235, 9, 2001.

[84]   S. Page and M. Coxon, "Virtual Reality Exposure Therapy for Anxiety Disorders: Small Samples and No Controls?," *Frontiers in Psychology*, vol. 7, 3, 2016. https://doi.org/10.3389/fpsyg.2016.00326

[85]   N. Nascivera, Y. M. Alfano, T. Annunziato, M. Messina, V. S. Iorio, V. Cioffi, R. Sperandeo, M. Rosato, T. Longobardi and N. M. Maldonato, "Virtual Empathy: The added value of Virtual Reality in Psychotherapy," in *2018 9th IEEE International Conference on Cognitive Infocommunications (CogInfoCom)*, 2018.

[86]   A. Rizzo, J. Difede, B. O. Rothbaum, J. M. Daughtry and G. Reger, "Virtual Reality as a Tool for Delivering PTSD Exposure Therapy," in *Post-Traumatic Stress Disorder: Future Directions in Prevention, Diagnosis, and Treatment*, Springer, 2013.

[87]   G. Riva, "Virtual Reality in Psychotherapy: Review," *CyberPsychology & Behavior*, vol. 8, pp. 220–230, 6, 2005.

[88] R. Giuseppe, B. Monica, B. Margherita, R. Silvia and M. Enrico, "Experiential Cognitive Therapy: A VR Based Approach for the Assessment and Treatment of Eating Disorders," *Studies in Health Technology and Informatics*, vol. 58, pp. 120–135, 1998.

[89] F. Vincelli, Y. H. Choi, E. Molinari, B. K. Wiederhold and G. Riva, "Experiential Cognitive Therapy for the Treatment of Panic Disorder With Agoraphobia: Definition of a Clinical Protocol," *CyberPsychology & Behavior*, vol. 3, pp. 375–385, 6, 2000.

[90] G. Riva, M. Bacchetta, M. Baruffi and E. Molinari, "Virtual Reality–Based Multidimensional Therapy for the Treatment of Body Image Disturbances in Obesity: A Controlled Study," *CyberPsychology & Behavior*, vol. 4, pp. 511–526, 8, 2001.

[91] G. Riva, M. Bacchetta, M. Baruffi and E. Molinari, "Virtual-Reality-based Multidimensional Therapy for the Treatment of Body Image Disturbances in Binge Eating Disorders: A Preliminary Controlled Study," *IEEE Transactions on Information Technology in Biomedicine*, vol. 6, pp. 224–234, 9, 2002.

[92] G. Riva, M. Bacchetta, G. Cesa, S. Conti and E. Molinari, "Six-Month Follow-Up of In-Patient Experiential Cognitive Therapy for Binge Eating Disorders," *CyberPsychology & Behavior*, vol. 6, pp. 251–258, 6, 2003.

[93] P. M. G. Emmelkamp, M. Bruynzeel, L. Drost and C. A. P. G. van der Mast, "Virtual Reality Treatment in Acrophobia: A Comparison with Exposure in Vivo," *CyberPsychology & Behavior*, vol. 4, pp. 335–339, 6, 2001.

[94] P. M. G. Emmelkamp, M. Krijn, A. M. Hulsbosch, S. de Vries, M. J. Schuemie and C. A. P. G. van der Mast, "Virtual Reality Treatment Versus Exposure In Vivo: A Comparative Evaluation In Acrophobia," *Behaviour Research and Therapy*, vol. 40, pp. 509–516, 5, 2002.

[95] N. Maltby, I. Kirsch, M. Mayers and G. J. Allen, "Virtual Reality Exposure Therapy for the Treatment of Fear of Flying: A Controlled Investigation," *Journal of Consulting and Clinical Psychology*, vol. 70, pp. 1112–1118, 10, 2002.

[96] B. K. Wiederhold, D. P. Jang, S. I. Kim and M. D. Wiederhold, "Physiological Monitoring as an Objective Tool in Virtual Reality Therapy," *CyberPsychology & Behavior*, vol. 5, pp. 77–82, 2, 2002.

[97] A. A. Benbow and P. L. Anderson, "A Meta-analytic Examination of Attrition in Virtual Reality Exposure Therapy For Anxiety Disorders," *Journal of Anxiety Disorders*, vol. 61, pp. 18–26, 1, 2019.

[98] R. A. McCann, C. M. Armstrong, N. A. Skopp, A. Edwards-Stewart, D. J. Smolenski, J. D. June, M. Metzger-Abamukong and G. M. Reger, "Virtual Reality Exposure Therapy for the Treatment of Anxiety Disorders: An Evaluation of Research Quality," *Journal of Anxiety Disorders*, vol. 28, pp. 625–631, 8, 2014.

[99] O. D. Kothgassner and A. Felnhofer, "Lack of Research On Efficacy Of Virtual Reality Exposure Therapy (VRET) for Anxiety Disorders in Children and Adolescents," *neuropsychiatrie*, no. 5, 2020. https://doi.org/10.3389/fpsyg.2016.00326

[100] F. Bonato, A. Bubka, S. Palmisano, D. Phillip and G. Moreno, "Vection Change Exacerbates Simulator Sickness in Virtual Environments," *Presence: Teleoperators and Virtual Environments*, vol. 17, pp. 283–292, 6, 2008.

[101] K. Nesbitt and E. Nalivaiko, "Cybersickness," in *Encyclopedia of Computer Graphics and Games*, Springer International Publishing, 2018, pp. 1–6.

[102] B. Keshavarz, R. Saryazdi, J. L. Campos and J. F. Golding, "Introducing the VIMSSQ: Measuring Susceptibility to Visually Induced Motion Sickness," *Proceedings of the Human Factors and Ergonomics Society Annual Meeting*, vol. 63, no. 1, pp. 2267–2271, 2019.

[103]   L. Rebenitsch and C. Owen, "Estimating Cybersickness from Virtual Reality Applications," *Virtual Reality*, no. 25, pp. 165–174, 2020.

[104]   M. Plusquellec, "Les mondes virtuels menacent-ils la santé mentale des enfants et des adolescents?," *Archives de Pédiatrie*, vol. 7, pp. 209–210, 2, 2000.

[105]   T. T. Holmberg, T. L. Eriksen, R. Petersen, N. N. R. Frederiksen, U. Damgaard-Sørensen and M. B. Lichtenstein, "Social Anxiety Can Be Triggered by 360-Degree Videos in Virtual Reality: A Pilot Study Exploring Fear of Shopping," *Cyberpsychology, Behavior, and Social Networking*, vol. 23, pp. 495–499, 7 July, 2020.

[106]   C. L. Mathews, H. E. R. Morrell and J. E. Molle, "Video Game Addiction, ADHD Symptomatology, and Video Game Reinforcement," *The American Journal of Drug and Alcohol Abuse*, vol. 45, pp. 67–76, 6, 2018.

[107]   F. Naeem, P. Phiri, T. Munshi, S. Rathod, M. Ayub, M. Gobbi and D. Kingdon, "Using Cognitive Behaviour Therapy with South Asian Muslims: Findings from the Culturally Sensitive CBT Project," *International Review of Psychiatry*, vol. 27, pp. 233–246, 2015.

[108]   N. Kumar and P. Gupta, "Cognitive Behaviour Therapy in India: Adaptations, Beliefs and Challenges," in F. Naeem and D. Kingdon (Eds.), *CBT in Non-Western Cultures*, Nova Publishers Inc, 2012.

[109]   *Mental Health Atlas 2017*, Geneva, Switzerland: World Health Organization, 2018.

[110]   G. M. Reger, P. Koenen-Woods, K. Zetocha, D. J. Smolenski, K. M. Holloway, B. O. Rothbaum, J. Difede, A. A. Rizzo, A. Edwards-Stewart, N. A. Skopp, M. Mishkind, M. A. Reger and G. A. Gahm, "Randomized Controlled Trial of Prolonged Exposure Using Imaginal Exposure Vs. Virtual Reality Exposure in Active Duty Soldiers With Deployment-related Posttraumatic Stress Disorder (PTSD)," *Journal of Consulting and Clinical Psychology*, vol. 84, pp. 946–959, 11, 2016.

2 Role of Swarm Intelligence and Neural Network in Intelligent Traffic Management

Umesh Kumar Lilhore and Sarita Simaiya
Chitkara University Institute of Engineering and Technology,
Chitkara University, Punjab, India

2.1 INTRODUCTION

Swarm intelligence (SI) explains the social structure of distributed, self-organized structures and it is the method in which a few distributed intelligent autonomous components operate and assist one another in the assignment. SI provides a framework that is important to investigate group (or dispersed) issues attempting to solve without centralization or even the provision of a modeling framework. The Internet of Things phenomenon, invented by Kevin Ashton, is a way of interacting that is increasingly gaining popularity across various fields of responsibility. This chapter provides a comprehensive analysis of the current intelligent transportation systems (ITM) as well as deep learning findings. The highway is indeed an often-ignored form of contemporary infrastructure investment. Researchers have all learned about self-driving vehicles, mapping applications and ride-hailing services. So, as it points out, a path itself can become a forum for an enormous variety of inventions [1].

Roadways can be updated through connectivity, illumination and power distribution technology that helps sustainable development, increase security and change the experience of driving. Roadways are no more a means to ride from one location to the next [2]. They can be used to charge electric vehicles as well as absorb solar power due to ITM's broad specific surface area. New technologies seem to be essential to keeping parts of the highways properly equipped with energy-efficient and environmentally safe technologies and equipment. Eventually, there'll be more innovation during the coming decades to build the highways smarter as well as more secure to drive. This study describes the development of smart traffic control focused upon this Internet of Things [3].

Intelligent transportation technology is storming the planet. Potential customers have a little more travel experience than most and different means of getting and using transportation facilities. Corporations have access to various international transportation including route optimization sectors. State legislators have different ideas to strengthen renewable, effective forms of transport to ease traffic as well as improve quality of life. In particular, the term IoT (Internet of Things) corresponds to the increasing range of electronic machines; millions of these systems may interact and connect towards others throughout the universe over the Internet and therefore can be monitored remotely and managed. The IoT also contains advanced systems and some other appliances. Besides instance, climate information is gathered only at the organizational level of IoT. IoT opens up new ways for communities to use data to monitor traffic, reduce pollution, enable efficient use of resources and keep the masses neat and healthy [4,5].

2.2   RELATED WORK

A few of the optimization algorithms for the transport network management proposed by researchers is categorized below, including their capacity and limitations.

In the research paper [6], the swarm intelligence techniques are mainly used to solve complex urban transportation issues and management processes. In reality, the consciousness of the insect species is premised on relatively basic instructions of personal insect actions. Among the different colony insects, the ant colony method succeeds in searching for food by taking the path with the largest pheromone amount collected by many other ant colonies. The pheromone current model is the messaging service between specific ants. Everything leads to the development of collective knowledge of socioeconomic ant colonies that can be deemed multi-agent technologies.

In the research paper [7], a fuzzy logic popped up in 1965 by Zadeh, implementing the theory of fuzzy sets. This was shown to be a really capable numerical approach for having to deal with objective reality, ambiguity, complexity and inaccuracy. Fuzzy logic is still being used as a conceptual framework to solve transportation complications like various traffic issues, accident investigation and preventative measures, speed enforcement at motorway junctions and traffic signal regulation. In the last couple of years, a few other innovations in data procurement innovations through innovative travel knowledge management have been completed. Few, however, cultural differences (like estimated travel time, make the journey distance, the regular driving speed of a car, climate data, individual tastes, road maintenance data, as well as other data which could be accessible to the control systems throughout actual) boost the ambiguity of the schedule selection [8].

In the research paper [9], the traditional minimum shortest path methodology like priority queues, Dijkstra shortest path algorithm and bidirectional search method are used by many investigations for traffic management. Similar to the difficulty of traffic management on system developments, it becomes much more complex to synchronize the behavior of a huge proportion of diverse traffic control instruments that are present on the Internet. Yet another way to handle this uncertainty is to split the cooperation text into subproblems and cohesive subproblems that can be fixed with a minimal level of communication. Multi-agent systems technologies can aid throughout the allocation of the issue as well as enable the cohesion of the operations of such operatives if needed.

In the research paper [10], a genetic algorithm (GA) has been demonstrated to be a successful optimization technique. It has been applied successfully to structural analysis or a vast variety of implementations such as channel estimation, planning, forwarding, regulation and many others. For implementing GAs to solutions to complex problems, has been the elevated computational provided professional to their predictable convergence rate is among the biggest barriers.

In the research paper [11], a multi-agent scheme in the first step was designed where each agent represents a vehicle that adapts to its requirements. The path to the congestion status of the real-time road network is through a two-objective process for optimization: the shortest path and the shortest path. The minimum congested degree of the connection is the goal. The representative-based method captures nonlinear vehicle feedback Routing activities and conditions of road-network congestion. Next, a series of quantitative indices have been designed to explain the congested degree of nodes, and such indexes have been used as the weights in the two-objective functions used: he routing and congestion reduction officers.

In the research paper [12], almost all of the time, ant colony optimization was used to fix transport issues like the traveling salesman problem (TSP) including vehicle routing (VRP) and a few works based on swarm intelligence are built to solve the problem of road traffic control. The problem cannot be solved with the classic versions: artificial ants just capable of producing successively shorter and more feasible tours using knowledge acquired in the form of a pheromone trail on the edges of the graph.

2.3   SWARM INTELLIGENCE, IOT AND NEURAL NETWORK IN ITM

AI-based swarm intelligence, IoT and neural network–based deep learning are mainly covered. An embedded system involvement utilizing swarm intelligence captures images and videos of the traffic on the road and statistical analysis of their behavior according to the surrounding environment and reroutes the traffic in various routes so congestion for each path will be reduced. The IoT innovation framework is like a collection of technologies, specifications and frameworks that proceed through simple visual Internet connections to the simplest and perhaps most complicated programs that use these smart devices, with the information they collect and interact as well as to the energy of the necessary stage to operate all such implementations [13,14].

The Internet of Things (IoT) seems to be a network of integrated and interdependent networks, computers, objects, individuals containing unique identifiers and the ability to transfer as well as exchange data across a network but without a human to human as well as the human to computer communication. [15]

By filling the gaps between the digital and physical worlds, IoT strives to create smart environments wherein families and communities will indeed be capable of living in a better and much more relaxed way. Even with this IoT application framework, there will be no way to do something with IoT devices without linking stuff to the network. Thus, simply stated, the IoT technology stack includes all the innovations required to transfer from the IoT system and data to the actual intent and function but for the so-often-called IoT application case [16].

2.4 SWARM INTELLIGENCE (SI)–BASED METHODS

This chapter summarizes quite a few SI-based methodologies, illustrating their remarkable variations, their importance and demerit points and their applications and services [17–23].

- **Genetic Algorithm (GA)** A genetic algorithm (GA) was originally proposed by Holland in 1975, and is indeed a local search optimization methodology premised on the dynamics of natural selection. The general premise of this method is to imitate the notion of the "fittest"; it replicates the procedures reported in a biological process where the powerful intends to evolve and sustain whereas the powerless appears to die. GA is a total population-oriented method wherein elements of the sample are levels of high density on one optimization technique's fitness [17].
- **Ant Colony Optimization (ACO)** Ant colony optimization is an evolutionary algorithm method inspired by Marco Dorigo's 1992 ant system (AS). This is influenced by the foraging behavior of real ants. The whole method consists of four major parts (ant, pheromone, daemon action as well as distributed control which contribute to the overall system. Ants are fictional agents that are used to imitate the discovery and manipulation of the search space. In real life, a pheromone is a chemical substance spread by ants along the road they move and its strength varies over time due to evaporation. In ACO, ants drop pheromones while moving in the search room [18].
- **Particle Swarm Optimization (PSO)** Particle swarm optimization (PSO) was presented by Kennedy and Eberhart in 1995 as a method for optimization [24]. The PSO method operates by initializing the population first, using an easy mechanism that simulates swarm behavior in birds wising up and also in social insects. The second phase includes calculating every particle's function value and revising the individual and worldwide bests as well as updating the particle speed and acceleration later. The rest of the penultimate stages are reiterated up to the time of dismissal [19].
- **Differential Evolution (DE)** The differential evolution (DE) method is a population GA-like methodology because it has comparable technicians: crossover, mutation as well as selection. This same key difference between DE and GA is whether DE is based on mutation while GA depends on crossover operations. Introducing this algorithm in 1997 can have as well as cost. While this method is reliant on the operation of mutation, it uses the mutation to search for potential regions in the search area and benefits from the analytical modeling [20].
- **Artificial Bee Colony (ABC)** The ABC is one of swarm intelligence's newest algorithms. In 2005, Dervis Karaboga proposed ABC[76], in 2007

ABC performance analyses were performed and in comparison with several other approaches, ABC was found to perform very well. The intelligent behavior of real honeybees searching for food sources called nectar and sharing information on this source of food amongst other bees in the nest inspired this algorithm. This is a simple and easily implemented algorithm, such as PSO and DE. This approach defines and categorizes the artificial agents in three different kinds: the bee employed, the bee viewers and the bee scout [21].

- **Glowworm Swarm Optimization (GSO)** The new SI technology for optimizing multi-modal operations is glow worm swarm optimization (GSO), suggested in 2005 by Krishnanad and Ghose. GSO uses glowworms called physical entities (accents). By taking into account the following amendments, GSO can generally be improved. (1) Extension to all agents of the neighborhood. When the best outcome is established, all operatives can move with the optimal method towards the agent. The whole step can boost operational efficiency as more agents are in the best solution range. (2) This same quantity of neighbors inside the community target area boosts the convergence speed for GSOs [22].

- **Cuckoo Search Algorithm (CSA)** Another of the recently launched metaheuristic approaches presented by Yang and Deb in 2009 is the cuckoo search algorithm (CSA). This concept is derived from the behavior and characteristics of Lévy flights of birds as well as fruit flows of cuckoo species like brood parasitic infections. In its execution, CSA uses three basic rules or operational activities. Initially, one egg can be laid in each incarnation for each cuckoo and the nest is randomly selected by the cuckoo for laying its egg. Secondly, eggs and nests are transported to the next generation with high quality. Thirdly, a bishop with probability discovers the number of host nests available, and the egg laid by a cuckoo [23].

2.5 ELEMENT OF IOT TECHNOLOGY

This can tend to be a difficult task if you try to navigate a path through all of the IoT application dungeons, considering the complexity as well as multiple regression analyses of technologies that comprise it. After all, in terms of ease, it can decompose the IoT technology platform across four simple forms of technology needed in making an Internet of Things functional [25–27]. These forms of technology are the following:

- **Device Hardware:** Machines are entities that necessarily imply "things" mostly on the Internet. Working as an intermediary between the actual and the virtual, these can have various sizes, shapes and degrees of technical sophistication describe the role they are motivated to deliver throughout the sense of a particular IoT infrastructure. Cameras, sensors and perhaps other sensor data equipment may also be stand-alone connected phones on their own. Just one restriction is found and the specific IoT uses scenarios as well as ITM hardware platforms width, simplicity of deployment and implementation, performance, life cycle, cost and effectiveness [25].

- **Device Software:** That's what makes the consumer devices intelligent. A model is responsible for incorporating interaction with both the Internet, collecting information, connecting applications and conducting real-time data analysis mostly on the IoT system. That's far more; it seems to be conducting training, which also offers input validation functionality for consumers to display information and transmit also with IoT infrastructure.

- **Communication:** Providing technology and software throughout location, there is another level that can provide connected devices through methods of sharing information with team members of the IoT network. Even though it is true which interaction processes are closely linked to software and hardware components devices, it really is important to regard these as distinct processes. The core network involves all external communication services (mobile, cellular, LAN) including basic standards used during various IoT systems (Thread, ZigBee, MQTT, LwM2). Picking the right advantage of those opportunities is among the main components of any IoT application framework [26].

- **Platforms:** As described earlier, according to the "smart" devices and technology installed, the system can "feel" what's going on in them and communicating this towards the consumer through a special interaction channel. IoT system architecture, durability, configuration features, standards used, technology agnosticism and safety as well as cost-effectiveness. It is therefore important to keep in mind that systems can be deployed on-site, including cloud-based. IoT system management software is a prime example of a framework that can be implemented both on-site and in the network. And the same relates to the other IoT network of the AV system [27].

2.6 IOT ARCHITECTURE

An extremely wide variety of linked heterogeneous networks and devices have been rendered, developing a general architecture for further IoT and quite a complex activity. Conversely, the IoT framework can usually be divided into three distinct layers, including the awareness layer also commonly referred to as the recognition layer. The perception layer will be accountable for collecting certain forms of data from the real environment and utilizing real end devices, such as RFID and readers, sensors, readers, GPS (Global Positioning System) receivers and all types of measurements [28]. A network layer, that is in the center, includes a variety of communications infrastructures that act as connection networks. This system is also responsible for data assortment, preliminary filtering data transmission. The top layer seems to be the application layer, which offers opportunities for business areas and new types of personalized solutions for specific customers. An application layer interface involves enhanced security authentication protocol, application restriction protocol, advanced message queue protocol, transport message queue protocol, expandable communication and presence protocol, an application delivery Service. Members can view IoT through mobile, personal computers, tablets or cell phones

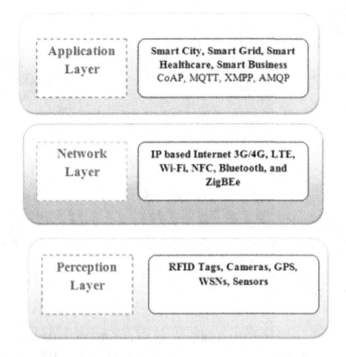

FIGURE 2.1 Architecture of IoT.

using the application layer framework. Depending on the service, certain devices such as intelligent refrigerators, mobile phones, smart TVs, etc. may also be used [29]. Figure 2.1 demonstrates the IoT architecture.

2.7 DEEP LEARNING

This chapter provides a comprehensive analysis of the current intelligent transportation systems (ITM) as well as the role of swarm intelligence and neural network (deep learning) findings. Deep learning (DL) is a conventional machine learning subsection where multiple layers of layered specifications are often used for learning. Such variables are element descriptions of specific facets that can influence the outcome of the system. There are several multilayer perceptions across each layer (often known as neurons) carrying loads for the variable. A component of such variables multiplies the input for every layer, and thus the outcome is a description of the effects in this input for each variable. A nonlinear system operates, generally after each layer or even several layers upon layers of neurons; for generating the final output, this same sigmoid is used. These same layers come together to form a deep layer called a DNNeep (Deep neural network) [30].

The back-propagation method is the most effective way of training the weight with variables in a centralized way for the final challenge. It is possible to find more details about this technique. Even though all the technologies that will also be

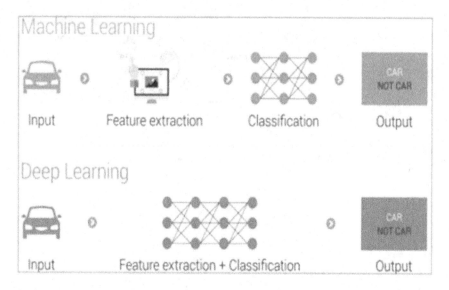

FIGURE 2.2   Deep Learning.

discussed in the remainder of this chapter can be classified as a DNN subset, DNN is the simplest network structure, i.e., fully connected layers. There is a link between all the neurons with one layer and all the neurons with one layer in this fully linked model. In some other layer, these same neurons, for each link, there seems to be a weight that should be dictated through the technique of back-propagation. Figure 2.2 shows the workings of the deep learning process.

2.7.1   TYPES OF DEEP LEARNING METHODS

In one of four fundamental communication networks, deep learning can be described as a neural network with such a large number of variables as well as layers [31–34]:

- **Convolutional Neural Network** A convolutional neural network (CNN also called ConvNet) is a deep neural network in deep learning, generally applied to visual imagery analysis. Based on the shared-weights architecture and transcription resemblance features, they are also known as change derivatives rather than space invariant artificial neural networks (or SIANN). Individuals who possess image data that allow analysts, recommendation system classification of images, medical imaging, and development of natural language as well as economic series data [31].
- **Recurrent Neural Network** A recurrent neural network (RNN) seems to be a neural network in which a directed graph across a temporal sequence forms connected nodes. It also enables the spatial dynamic model to be exhibited. RNNs might use one's inner state recollection to handle the length in patterns of components obtained from recurrent neural

networks. The latter makes others relevant to assignments such as un-regimented, linked acknowledgment of handwritten or acknowledgment of talk [32].

- **Unsupervised Pre-trained Network** Unsupervised pre-training initializes a discriminative neural net, including a deep convolutional neural network or even deep autoencoders, from one that was prepared to use an un-supervised set of criteria. The whole technique sometimes can assist both with the problems of system integration as well as generalization [33].
- **Recursive Neural Network** A recursive neural network is a type of deep neural network generated by remaking the very same weight matrix over just a highly skilled, by crossing a harding in geometrical order to create an organized forecast over variable-size feature formed or even a scalar es-timation upon this. Throughout natural language processing, for example, recursive neural providers, occasionally referred to as RvNNs, have also been confident in teaching pattern and forest structures, primarily term and word constant depictions based on deep learning. To understand dis-persed structure depictions, like logical terms, RvNNs first were presented. Although the 1990s, designs and general implementations in any further works are being established [34].

2.8   ADVANTAGES OF AN ITM

ITMs can be used in parking areas, speed cameras, traffic signals, highways and bridges that are used to develop interrelated transport networks of fully accessible communication of data as well as vehicles [35].

Major ITM benefits:
- Development of interconnected transport systems with open interaction around devices as well as transport vehicles.
- Consciously manage traffic, working to maintain transportation.
- Trying to ensure residents have access to real-time vehicles and mass transit details. It also decreases travel distances for passengers as well as tends to make transportation around the metropolitan area simpler, secure and much more convenient.
- ITM can properly monitor traffic such that it goes smoothly and perhaps public transportation finally arrives as originally scheduled.
- ITM guarantees that individuals possess access to real-time traffic and mass transit data.

2.9   KEY APPLICATIONS OF ITM

ITM application areas include leverage sensing data, such as GPS trackers and highway camera systems. Below are five main uses that smart cities are adopting as a component of their ITM frameworks [36–39].

- **Advanced Traffic Management System:** This primarily handles traffic by integrating data from a variety of channels, including traffic signals, toll roads as well as parking spaces. All such components offer revised data on the status of such traffic flow. ATMS encourages proper functioning with vehicles, which also helps to lower the emissions standards of the metropolis. ATMS includes monitoring major roads as well as giving guidelines to vehicles for the secure, quicker experience of driving [36].
  ATMS can greatly reduce traffic and increase performance:
  - **Configure traffic lights as well as other alerts** –Throughout real time, attempting to shift viewers away from overcrowded routes as well as on traffic-free roads.
  - **Automatically configure the frequency on tollways** – Dissuade drivers from using personal vehicles and also to promote all use of public buses.
  - **Keep providing traffic data** – Warn drivers to park slot sections, improve efficiency as well as end up searching for parking facilities.
- **Advanced Public Transportation System (APTS):** This primarily gives people access to data on bus routes, such as available slots, destination and arrival time. Everything just tends to make payment smoother and much more responsive by permitting the use of smartphone bookings or "near field communication" to reimburse for numerous forms of transport. This same framework can often end up making dynamic decisions, for example, delaying public buses attempting to run sooner than expected [38].
  To maintain the effectiveness of public transport, APTS seems to use:
  - The real-time travelers' data system
  - Automated traffic positioning devices
  - Vehicle travel update device
  - Devices giving primary responsibility for bus routes at junctions
- **Commercial Vehicle Operation (CVO):** These are primarily used in passenger vehicles, which include public transport, ambulance crews, trucks as well as cabs. The CVO also provides automated traffic tracking. The whole system is an electronic aspect used to track performance as well as driving behavior. This is essential because poor behavior mostly on the highway can cause injuries. Another other model is fleet control, which also enables the effective distribution of automobiles such as smartphones, GPS, sensors, etc. By using the information gathered, a corporation can evaluate ITM's own fleet to minimize cost. This can be achieved by monitoring fuel consumption, verifying driver adherence most with new routes and security protocols, analyzing the data required for operating expenses, etc. All such methodologies are being used to improve communications among regulatory bodies and operators, cut costs as well as facilitate the effective transition of various goods and services [39].

Other elements of the CVO must therefore usually involve:
- Presidency of freight
- Heavy vehicle regulatory method
- Mass-in-motion
- Side of the road CVO security
- Tracking of the on-board protection
- Hazardous material starting to plan as well as reaction
- **Advanced Vehicle Control and Safety System (AVCSS):** It is primarily based on sensors installed throughout automobiles. Individuals are presented with graphic notifications as well as details on potential hazards. Until the moment arises that if automobiles are highly automated, AVCSS also provides technological innovation that increases and facilitates automated cars. AVCSS-equipped automobiles may be regarded as an independent tier 1 device. This same SAE J3016 201401 specification specifies five categories of automated vehicles, in which Level 0 is a device with no automation, as well as Level 5 is completely automated, capable of driving even under the road but also environmental problems without the need of a car driver [37].

AVCSS can enhance traffic safety:
- **Automatically adjust the automobile,** reduce or prevent manual driving mistakes.**Help people** minimize as well as quickly respond to dangerous conditions.
- **Focus on providing** cruise control, traffic maintenance as well as speed changes.

2.10   INTELLIGENT TRANSPORT SYSTEM (ITM) FUNCTIONALITIES

Everything mainly utilizes multiple techniques to increase traffic flow. An intelligent system of interconnected sensors as well as connected phones can open the way for a much more sustainable, efficient and robust center of the city. In particular, this same ITM will set the stage for a futuristic world that can be quickly crossed by a blend of music and new transit options. Because once put into effect and widely used by investors, ITM technologies can improve mass transit flow to a level at which motor vehicles will not be required and a need for parking space as well as four-lane highways have been significantly lowered. Roadways filled with cars, petrol as well as overcrowding may have been a regular occurrence [24,40,41]. The ITM can adopt the following basic guidelines:

- **Collection of Traffic Data:** covers all the major applications focus on data compilation. Sensors, as well as dynamic technologies, obtain real-time information, which includes destination and vehicle speeds, the timing of difficulties as well as the speed limit. Several of the equipment that is used to collect the information seems to be: GPS, sensors, cameras on road, system for vehicle identifiers and signal data.
- **Data Transmission:** The data obtained is transferred from the sensor detectors to the monitoring center. These results are recorded as well as

submitted to application fields just at the operations center. Relevant evidence is then shared with residents.

- **Traffic Data Analysis:** Data is extracted, cleaned and customized in the control center for further evaluation. It's also consolidated so that monitoring and controlling as well as forecasts can be decided to make before being sent to end-user functionalities.
- **Traveler Information:** Traveler information in an order made publicly available and used by residents. This can be obtained through the platforms, such as broadcast, the Internet as well as notifications. All of the information assures residents of existing traffic patterns, disruptions, highway circumstances, speed and time.

2.11   CHALLENGES BESIDES TRANSPORT IN METROPOLITAN CENTERS

The following are the recommended transport issues that need to be addressed:

- **Traffic Congestion Issues:** Overloading seems to be the main cause of overcrowding. The property utilizes trends and transportation infrastructure affects the speed of traffic. Even though both travelers, as well as cargo traffic, make a significant contribution to traffic jams, the moves of passengers are also the primary source of deadlock in metropolitan centers. Throughout the 21st century, road users spent four times longer in traffic than even the drivers did several generations earlier. Massive groups of single-seat automobiles contribute to the flow of vehicles. The consequent overcrowding contributes significantly to environmental pollution, ineffective use of energy and weaker travel, making the city landscape annoying. Drivers face challenges such as buses, delivery vans and utility vehicles, looking for parking spots near their target [42].
- **Long Commutes:** Increasing trends, highway construction as well as travel between offices and homes lead to an increase in traffic and longer service periods. The expansion of traffic density may not always be effective in reducing travel times, since it is not keeping pace with the increased level of traffic. In reality, new infrastructure will lead to stronger journeys, as they allow further traffic to use transport infrastructure and increase the average vehicle-mile travel journey. Residence accessibility frequently influences the behaviors of traveling. Whereas most job prospects exist in metropolitan areas, suburban accommodation seems to be more desirable. But affordable accommodation comes at the price of longer traveling costs [43].
- **Sprawling Cities:** Decentralization also had the effect of increasing transportation systems and are complicated. When populations expand geographically, and the gap between residential buildings and places of employment rises, traffic becomes a significant obstacle for societies, and transportation times become a massive burden for individual citizens. Urban development allows public transit networks to become more costly

to construct and maintain and limits the mobility of pedestrians. Supermarket chains as well as other services servicing wide transport links aren't openly available on feet, and this promotes further use of automobiles [44].

- **Secondary Infrastructure:** Interest in vehicular and pedestrian facilities is continuing to grow since more users enjoy riding and biking to work. Even so, several cities have been designed for vehicles and are therefore not bicycle-friendly and pedestrian-friendly. Cycle paths, including wider sidewalks, make cycling and walking perfectly safe and therefore can help regulate accidents, but this maintenance comes at the price of traffic congestion and parking areas. Public transport also needs parking facilities. Suburban locations may provide cycle lanes for travelers to encourage the use of public transportation. Passengers may use these suburban platforms to avoid the embarrassment of parking costs [45].

- **Large Fleets, Large Costs:** Urban service providers experience difficulties in maintaining huge automobiles and increasing workforces, including infrastructure costs, attraction and training of qualified staff and functioning correctly for tasks. Such organizations should therefore instruct their employees to improve security and reduce casualties. Rapidly changing the level of public transportation is a challenge for public travel agencies, which must change the value of each fleet. The fleet that is massive enough though to satisfy peak-hour competition isn't financially viable while runoff anyway-peak. Even so, if carriers do not have the cars, they can't fully accommodate the number of customers during peak times [46].

- **Parking Difficulties:** Vehicles stuck in a traffic jam when searching for a parking space add to traffic congestion. Urban areas are struggling to have adequate parking areas to support centralized commercial areas. Wide parking spaces use costly rental properties, whereas a parking space uses up roads that can be used to transfer vehicles [47].

- **Negative Environmental Impacts:** The dependence on automobiles influences the health and well-being of people, particularly the population's health. Automobiles including support products have impacts on the environment in urban centers. Pollution levels, particularly carbon dioxide emissions, increase throughout traveling automobiles. Road networks account for between 30 and 60% of suburban land, although their strategic impact grows because more people are using personal vehicles. Traffic produces disturbances and smells that simply makes walking in densely populated urban areas unpleasant. Longer access to such fumes, especially if the motor is not properly managed, is harmful to health. Automotive fumes contain carbon dioxide, fatty acids, carbon monoxide as well as other gases in the atmosphere and contaminants such as swordtails-ethyl carbon, nitrous oxide as well as carbonaceous material [48,49].

2.12 CONCLUSIONS

This whole survey was particularly concerned only with actual implementation of a specific swarm intelligence (SI)–oriented approach as well as targeted to provide an application in ITM. A set of methodologies such as ant colony optimization (ACO), glow worm swarm optimization (GSO), genetic algorithm (GA), particle swarm optimization (PSO), artificial bee colony (ABC), differential evolution (DE), as well as cuckoo search algorithm (CSA) was discussed. The pace of technological change, i.e., AI, ML, cloud computing, IoT, is endless. This offers a solution throughout all sectors, including production, manufacturing, apparel, restaurants, healthcare and education. The intelligent transport systems-based "smart cities" can provide a popular green technology design that is appropriate, especially for small towns. The cloud-based design of IoT applications for smart city applications becomes suitable through utilizing an information-sharing framework. Metropolitan areas may make up a popular interconnected system. Throughout this way, technologies from small and large smart grids are connected and managed through the central cloud service. Last but not least, the scale of a region is not a major obstacle to being "smart." Intelligent technology will be used by communities across each category.

REFERENCES

[1] Bhardwaj, K.K., Khanna, A., Sharma, D.K., and Chhabra, A., 2019. Designing energy-efficient IoT-based intelligent transport system: Need, architecture, characteristics, challenges, and applications. In *Energy Conservation for IoT Devices* (pp. 209–233). Springer, Singapore.

[2] Levina, A.I., Dubgorn, A.S., and Iliashenko, O.Y., 2017, November. Internet of Things within the service architecture of intelligent transport systems. In *2017 European Conference on Electrical Engineering and Computer Science (EECS)* (pp. 351–355). IEEE.

[3] Geetha, S., and Cicilia, D., 2017, October. IoT Enabled intelligent bus transportation system. In *2017 2nd International Conference on Communication and Electronics Systems (ICCES)* (pp. 7–11). IEEE.

[4] Eswaraprasad, R., and Raja, L., 2017, December. Improved intelligent transport system for reliable traffic control management by adopting internet of things. In *2017 International Conference on Infocom Technologies and Unmanned Systems (Trends and Future Directions)(ICTUS)* (pp. 597–601). IEEE.

[5] Sodhro, A.H., Obaidat, M.S., Abbasi, Q.H., Pace, P., Pirbhulal, S., Fortino, G., Imran, M.A., and Qaraqe, M., 2019. Quality of service optimization in an IoT-driven intelligent transportation system. *IEEE Wireless Communications*, 26(6), pp. 10–17.

[6] Zambada, J., Quintero, R., Isijara, R., Galeana, R., and Santillan, L., 2015, October. An IoT-based scholar bus monitoring system. In *2015 IEEE First International Smart Cities Conference (ISC2)* (pp. 1–6). IEEE.

[7] Muthuramalingam, S., Bharathi, A., Gayathri, N., Sathiyaraj, R., and Balamurugan, B., 2019. IoT-based intelligent transportation system (IoT-ITM) for global perspective: A case study. In *Internet of Things and Big Data Analytics for Smart Generation* (pp. 279–300). Springer, Cham.

[8] Costantini, F., Archetti, E., Di Ciommo, F., and Ferencz, B., 2019. IoT, intelligent transport systems, and MaaS (mobility as a service). *Jusletter IT*, 21.

[9] Perallos, A., Hernandez-Jayo, U., Onieva, E., and Zuazola, I.J.G. eds., 2015. *Intelligent Transport Systems: Technologies and Applications*. John Wiley & Sons.

[10] Wang, D., Chen, D., Song, B., Guizani, N., Yu, X., and Du, X., 2018. From IoT to 5G I-IoT: The next generation IoT-based intelligent algorithms and 5G technologies. *IEEE Communications Magazine*, 56(10), pp. 114–120.

[11] Carignani, M., Ferrini, S., Petracca, M., Falcitelli, M., and Pagano, P., 2015, December. A prototype bridge between automotive and the IoT. In *2015 IEEE 2nd World Forum on the Internet of Things (WF-IoT)* (pp. 12–17). IEEE.

[12] Pattar, S., Sandhya, C.R., Vala, D., Buyya, R., Venugopal, K.R., Iyenger, S.S., and Patnaik, L.M., 2019, December. SoCo-ITM: Service-oriented context ontology for the intelligent transport system. In *Proceedings of the 2019 7th International Conference on Information Technology: IoT and Smart City* (pp. 503–508).

[13] Chand, H.V., and Karthikeyan, J., 2018. Survey on the role of IoT in intelligent transportation system. *Indonesian Journal of Electrical Engineering and Computer Science*, 11(3), pp. 936–941.

[14] Dado, M., Janota, A., Spalek, J., Holečko, P., Pirník, R., and Ambrosch, K.E., 2015, October. Internet of Things as advanced technology to support mobility and intelligent transport. In *International Internet of Things Summit* (pp. 99–106). Springer, Cham.

[15] Zhang, W.Y., Wang, X.F., and Feng, X., 2013. Research on model-designing & architecture of IoT-based intelligent transportation system. In *Applied Mechanics and Materials* (Vol. 392, pp. 991–996). Trans Tech Publications Ltd.

[16] Mogi, R., Nakayama, T., and Asaka, T., 2018, November. Load balancing method for IoT sensor system using multi-access edge computing. In *2018 Sixth International Symposium on Computing and Networking Workshops (CANDARW)* (pp. 75–78). IEEE.

[17] Finogeev, A., Finogeev, A., Fionova, L., Lyapin, A., and Lychagin, K.A., 2019. Intelligent monitoring system for smart road environment. *Journal of Industrial Information Integration*, 15, pp. 15–20.

[18] Ata, A., Khan, M.A., Abbas, S., Khan, M.S., and Ahmad, G., 2020. Adaptive IoT empowered smart road traffic congestion control system using supervised machine learning algorithm. *The Computer Journal*. https://doi.org/10.1093/comjnl/bxz129

[19] Khanna, A., and Anand, R., 2016, January. IoT-based smart parking system. In *2016 International Conference on Internet of Things and Applications (IOTA)* (pp. 266–270). IEEE.

[20] Misbahuddin, S., Zubairi, J.A., Saggaf, A., Basuni, J., Sulaiman, A., and Al-Sofi, A., 2015, December. IoT based dynamic road traffic management for smart cities. In *2015 12th International Conference on high capacity optical networks and enabling/emerging technologies (HONET)* (pp. 1–5). IEEE.

[21] Sharif, M., Mercelis, S., Van Den Bergh, W., and Hellinckx, P., 2017, December. Towards real-time smart road construction: Efficient process management through the implementation of the internet of things. In *Proceedings of the International Conference on Big Data and Internet of Thing* (pp. 174–180).

[22] Frank, A., Al Aamri, Y.S.K., and Zayegh, A., 2019, January. IoT-based smart traffic density control using image processing. In *2019 4th MEC International Conference on Big Data and Smart City (ICBDSC)* (pp. 1–4). IEEE.

[23] Masek, P., Masek, J., Frantik, P., Fujdiak, R., Ometov, A., Hosek, J., Andreev, S., Mlynek, P., and Misurec, J., 2016. A harmonized perspective on transportation management in smart cities: The novel IoT-driven environment for road traffic modeling. *Sensors*, 16(11), p. 1872.

[24] Devi, S., and Neetha, T., 2017. Machine Learning-based traffic congestion prediction in an IoT-based Smart City. *International Research Journal of Engineering and Technology*, 4, pp. 3442–3445.

[25] Prasada, P., and Prabhu, K.S., 2020. Novel approach in IoT-based smart road with traffic decongestion strategy for smart cities. In *Advances in Communication, Signal Processing, VLSI, and Embedded Systems* (pp. 195–202). Springer, Singapore.

[26] Mansoori, A., and Achar, C., 2018. Smart roads using IoT devices. *International Research Journal of Engineering and Technology*, 5(6), pp. 1526–1529.

[27] Al-Shammari, H.Q., Lawey, A., El-Gorashi, T., and Elmirghani, J.M., 2019, July. Energy-efficient service embedding in IoT over PON. In *2019 21st International Conference on Transparent Optical Networks (ICTON)* (pp. 1–5). IEEE.

[28] Zhao, L., Wang, J., Liu, J., and Kato, N., 2019. Optimal edge resource allocation in IoT-based smart cities. *IEEE Network*, 33(2), pp. 30–35.

[29] Janahan, S.K., Veeramanickam, M.R.M., Arun, S., Narayanan, K., Anandan, R., and Parvez, S.J., 2018. IoT-based smart traffic signal monitoring system using vehicle counts. *International Journal of Engineering & Technology*, 7(2.21), pp. 309–312.

[30] Keshavarzi, A. and van den Hoek, W., 2019. Edge intelligence—On the challenging road to a trillion smart connected IoT devices. *IEEE Design & Test*, 36(2), pp. 41–64.

[31] Zantalis, F., Koulouras, G., Karabetsos, S., and Kandris, D., 2019. A review of machine learning and IoT in smart transportation. *Future Internet*, 11(4), p. 94.

[32] Saarika, P.S., Sandhya, K., and Sudha, T., 2017, August. Smart transportation system using IoT. In *2017 International Conference On Smart Technologies For Smart Nation (SmartTechCon)* (pp. 1104–1107). IEEE.

[33] Babar, M., and Arif, F., 2019. Real-time data processing scheme using big data analytics on the internet of things-based smart transportation environment. *Journal of Ambient Intelligence and Humanized Computing*, 10(10), pp. 4167–4177.

[34] Rathore, M.M., Paul, A., Hong, W.H., Seo, H., Awan, I., and Saeed, S., 2018. Exploiting IoT and big data analytics: Defining smart digital city using real-time urban data. *Sustainable Cities and Society*, 40, pp. 600–610.

[35] Mrityunjaya, D.H., Kumar, N., Ali, S., and Kelagadi, H.M., 2017, February. Smart transportation. In *2017 International Conference on I-SMAC (IoT in Social, Mobile, Analytics, and Cloud)(I-SMAC)* (pp. 1–5). IEEE.

[36] Zantalis, F., Koulouras, G., Karabetsos, S., and Kandris, D., 2019. A review of machine learning and IoT in smart transportation. *Future Internet*, 11(4), p. 94.

[37] Mahdavinejad, M.S., Rezvan, M., Barekatain, M., Adibi, P., Barnaghi, P., and Sheth, A.P., 2018. Machine learning for Internet of Things data analysis: A survey. *Digital Communications and Networks*, 4(3), pp. 161–175.

[38] Jan, B., Farman, H., Khan, M., Talha, M., and Din, I.U., 2019. Designing a smart transportation system: An Internet of things and big data approach. *IEEE Wireless Communications*, 26(4), pp. 73–79.

[39] Bansal, K., Mittal, K., Ahuja, G., Singh, A., and Gill, S.S., 2020. DeepBus: Machine learning-based real time pothole detection system for smart transportation using IoT. *Internet Technology Letters*, 3(3), p. e156.

[40] Simaiya, S., Lilhore, U.K., Sharma, S.K., Gupta, K., and Baggan, V., 2020. Blockchain: A new technology to enhance data security and privacy in the Internet of Things. *Journal of Computational and Theoretical Nanoscience*, 17(6), pp. 2552–2556.

[41] Ma, X., Yao, T., Hu, M., Dong, Y., Liu, W., Wang, F., and Liu, J., 2019. A survey on deep learning empowered IoT applications. *IEEE Access*, 7, pp. 181721–181732.

[42] Babar, S., Mahalle, P., Stango, A., Prasad, N., and Prasad, R., 2010, July. The proposed security model and threat taxonomy for the Internet of Things (IoT). In *International Conference on Network Security and Applications* (pp. 420–429). Springer, Berlin, Heidelberg.

[43] Pawar, A.B., and Ghumbre, S., 2016, December. A survey on IoT applications, security challenges, and countermeasures. In *2016 International Conference on Computing, Analytics and Security Trends (CAST)* (pp. 294–299). IEEE.

[44] Miraz, M.H., and Ali, M., 2018, August. Blockchain-enabled enhanced IoT eco-system security. In *International Conference for Emerging Technologies in Computing* (pp. 38–46). Springer, Cham.

[45] Xiao, L., Wan, X., Lu, X., Zhang, Y., and Wu, D., 2018. IoT security techniques based on machine learning: How do IoT devices use AI to enhance security?. *IEEE Signal Processing Magazine*, 35(5), pp. 41–49.

[46] Chin, J., Callaghan, V., and Lam, I., 2017, June. Understanding and personalizing smart city services using machine learning, the internet-of-things, and big data. In *the 2017 IEEE 26th International Symposium on Industrial Electronics (ISIE)* (pp. 2050–2055). IEEE.

[47] Strohbach, M., Ziekow, H., Gazis, V., and Akiva, N., 2015. Towards a big data analytics framework for IoT and smart city applications. In *Modeling and processing for Next-generation Big-data Technologies* (pp. 257–282). Springer, Cham.

[48] Strohbach, M., Ziekow, H., Gazis, V., and Akiva, N., 2015. Towards a big data analytics framework for IoT and smart city applications. In *Modeling and processing for Next-generation Big-data Technologies* (pp. 257–282). Springer, Cham.

[49] Park, J.H., Salim, M.M., Jo, J.H., Sicato, J.C.S., Rathore, S., and Park, J.H., 2019. CIoT-Net: a scalable cognitive IoT-based smart city network architecture. *Human-centric Computing and Information Sciences*, 9(1), p. 29.

3 Image Systems and Visualizations

*Pulkit Narwal, Neelam Duhan
and Komal Kumar Bhatia*
J.C. Bose University of Science and Technology, YMCA,
Faridabad, Haryana, India

3.1 INTRODUCTION

The field of artificial intelligence has opened the gates of wide computational and logical abilities for the computer system. The basic idea behind this has been to mimic the human learning abilities and provide a mechanism that replicates the logical understanding and decision making of humans. This mechanism can be understood through a multiple number of theories. This can be explained by experience learning, where the humans learn from their previous experiences and the brain stores the prior steps, the process and the results of that process into an experience. This experience can be treated as an encapsulated unit of data. Similarly new experiments have shown that experience learning can be used by computers where historical data is fed to the system as experience input, which in turn processes the data or trains to obtain the results and thus stores the labelled output data as a result of the process.

Computers can use their learning to apply their knowledge that has been derived from the learning process towards solving new, similar problems. But, letting computers learn is itself a task. Learning can be either supervised where the data is labelled and computers can later match the new query with the labelled data and classify the results. An unsupervised way of learning is more inferential based, where certain patterns are extracted out of the data and the patterns are used to now cluster the data into different groups. Semi-supervised learning has also been a new idea, where it can act both as supervised as well as unsupervised learning based on whether the system has prior knowledge to the problem or not.

Recent explorations have also suggested that reinforcement learning is an ideal way to mimic the human experience and decision-making abilities. But due to its lack of range limitations, i.e., the environment has to be completely defined, it lacks some computational issues. Consider an example of a child; a child performs certain actions that may be considered as trial-and-error terminology. The child either finds something interesting in an activity, which may be considered a reward, and the child now stores this experience as a good one, one that gives him joy. The child may also find something that troubles the child or is less liked; be it falling down or hunger. So this experience can be stored as a bad experience or the one that gives a penalty. This reward and penalty mechanism depicts the nature of reinforcement learning. The reward and penalty rules are defined and the goal is to maximize the overall reward. Artificial intelligence can be termed as the intelligence in computers that resembles human intelligence.

Computer vision is a subfield of artificial intelligence that aims to provide intelligence relating to visual cognitive skills. The ability to see things and understand it and make decisions describes the art of computer vision. Humans receive visual input from the environment by rays reflected back from an object to their eyes. The retina receives all the rays and the brain performs the understanding procedure. Cameras are considered to be the eyes of the computer that receives the input rays from the environment via a camera lens. The visual scene can either be a static 2D object (image) or a dynamic collection of 2D images (video) containing the motion information as well. After the image acquisition, processing an image is the next step. Processing may be viewed as the perception of an image or visualization of the image; in computer vision, it can also be referred to as pre-processing. Based on the processing, decision making is performed. This decision making may involve the type of learning or rules that the computer system has been trained with. The below diagram depicts the human visual system and computer vision system (Figures 3.1, 3.2, and 3.3).

Computer vision is defined as the art of understanding the image replicating the human visual process. In other words, computer vision is the interpretation of the image or scene understanding from an image. Often the terms image processing, computer graphics and computer vision are misunderstood to be the same. Image processing is the process of image recovery, reconstruction, filtering, compression and visualization. The input and output states of image processing involve an image. Computer graphics is based on synthesis. Computer graphics involves synthesis of an image model into an image. Computer graphics create or synthesize

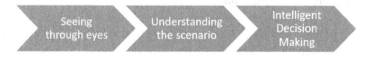

FIGURE 3.1   Human Visual System.

FIGURE 3.2   General Vision System.

FIGURE 3.3   Computer Vision System.

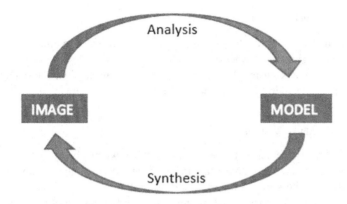

FIGURE 3.4   Image Analysis and Synthesis.

an image from a scene representation, whereas computer vision is simply the analysis of an image into an image model [1] (Figure 3.4).

Consider the Bayesian inference:

Probability of world with the image given, P (World/Image) can be defined as computer vision

$$P(\text{World/Image}) = (P(\text{Image/World}) * P(\text{World}))/P(\text{Image})$$

where P(Image/World) is accounted for computer graphics (probability of image given the world).

And, P(World) is the probability of the world and P(Image) is the probability of the image.

Computer vision depends on the following:

- Geometry
- Physics
- The nature of objects in the real world

3.1.1   Computer Vision Is an Ill-posed Inverse Problem

Biological and digital imaging process converts 3D real-world objects/scenes into 2D projections or images; the image is two dimensional (x-dimension and y-dimension). The z-dimension is lost in this conversion. On the contrary, computer vision is an inverse problem. Computer vision and image understanding is the 3D reconstruction of the two-dimensional image. Computer vision is an ill-posed problem because the depth information or the z-dimension is missing in the image. Determining depth from an image makes computer vision an ill-posed problem. Stereopsis or stereo vision, shape from shading and photometric stereo are some techniques that can approximate the depth information from image(s).

3.1.2 MID-LEVEL VISION

- Segmentation: Segmentation refers to partitioning of connected homo-geneous regions/components.
- Tracking: Tracking involves checking or verifying against a trajectory of object in consideration with the help of multiple cameras.

3.1.3 HIGH-LEVEL VISION

- Geometry: Understanding the relationship between object geometry and image geometry. The understanding includes finding the position and or-ientation of the known objects, determining the boundary of curved object and how the boundary or outline moves when the curved object is viewed from different directions.
- Probabilistic: Probabilistic vision uses classifiers and probability model to recognize different objects in the image.

3.2 FUNDAMENTALS OF COMPUTER VISION

Computer vision works on scene understanding and thus developing an interpretation of image [2], [1], [3]. Computer vision is an abstract process, which builds an understanding of the image by analysing and interpreting the image. This process can be segregated into different modules with each module having its significance. The fundamental steps include pre-processing (image enhancement and image reconstruction), segmentation (isolating objects), feature extraction (representation and description), classification (object recognition) and intelligent decision making (Figure 3.5).

3.2.1 PRE-PROCESSING

Image acquisition converts light reflected from objects into electrical signals to form an analog image. To perform operation on the image, the image conversion is required from an analog image to an digital image. An analog image depicts the continuous transition in intensities, whereas a digital image involves discrete transitions.

A digital image is a 2D area of numbers (quantized intensity values). Pre-processing emphasizes the better clarity of pictorial information. The image received is converted into a digital image using sampling along the x-axis and y-axis, called spatial sampling, followed by quantization of intensity values. The digital image is defined over a grid, where each grid location may be viewed as a pixel.

Types of digital images:

- Binary Image (Black and White): A binary image is formed of black and white pixels representing binary information (1 or 0). A binary image requires 1-bit pixel, which makes the binary image efficient in terms of storage. A binary image is used in document processing, OCR (Optical Character Recognition), contour(optical character recognition), contour representation, fingerprints, etc.

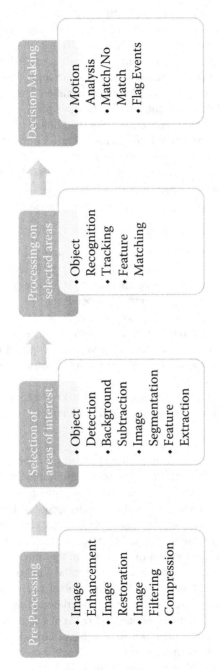

FIGURE 3.5 Computer Vision Steps [1].

- Grayscale Image: Grayscale image uses the intensity range in the power of 2. There are 256 intensity levels that are sufficient for most of the applications. A grayscale image requires 8 bits per pixel.
- RGB Image: An RGB image is also called a true color image, since it gives the actual color information of the image. RGB has three components namely: R (red) component, G (green) component, and B (blue) component. Each color component intensity gives a unique color when combined. Each pixel has three components. This pixel is called a vector pixel. An RGB image is generally represented by 24 bits per pixel.
- Indexed Image: An indexed image has a color map linked to the image that contains the color information in RGB format. Image information is stored in indices where each indices index represents a color map value or RGB value. An indexed image is generally used in GIF, PNG formats, etc. An indexed image is generally represented by 8 bits per pixel.

3.2.1.1   Image Enhancement

The digital image so formed may not have good visual quality. Image enhancement deals with improving the visual quality of image. Image enhancement may include increasing contrast, improving brightness, sharpening edges, removing edges, etc. Image enhancement is subjective in nature. It depends on the preference of observer. One may term an image good quality, which has been termed not good by the other observer. There is no mathematical theory defined, given the subjective nature of image enhancement. Image enhancement can be carried out either in a spatial domain where pixel intensities are manipulated directly or a frequency domain where the Fourier transform is modified.

Accomplishing image enhancement may require some operations based on their domain and application. Here are some spatial domain and frequency domain techniques:

i. Point Operation: Point operation technique works on the gray-level intensity values of the pixels itself. A point operation may be plotted on a graph with gray-level intensity on the x-axis and number of pixels on the y-axis. Transformations are then applied on the gray-level values. Point operation is the simplest case. This can also be referred to as brightness transform or point processing. Point operation depends only on the intensity value, not on the position of the pixel in the image.

ii. Spatial Operation: Spatial operation uses the spatial neighbourhood of the pixels. Pixels are treated as a grid. For given pixel coordinates, a pixel window is considered to ascertain the neighbourhood pixels. The neighbourhood can either be four-neighbourhood or eight-neighbourhood for a given 2D pixel grid. Operations are then applied, considering the point pixels as well as neighbourhood pixels.

iii. Frequency Domain Operations: Frequency domain works on frequency information of the image. At first, the Fourier Transform converts the given image into a frequency domain. Frequency level image enhancement operations are then applied to enhance the image. Then the enhanced image is converted back into spatial domain along with intensities. This back conversion is done through inverse Fourier Transform.

iv. Pseudo Colouring: Pseudo colouring, also known as false colouring, is the conversion of gray-scale image into color image.

3.2.1.2   Image Restoration

Image restoration works on restoring the lost feature of the image due to some factors. The factors may include not properly focussing of the object and motion blur resulting in blurry image or foggy image. The principle behind image restoration is objective in nature. Provided that a certain mathematical model can be applied to restore the image. De-blurring is the most frequently used image restoration technique.

3.2.1.3   Image Compression

Image compression is the technique of compressing the image data for better resource management, given the resources of storage space and transmission bandwidth. The idea is to convey a maximum by sending a minimum. The size should be less and the information conveyed should not be compromised. Such a pre-processing strategy is known as image compression.

3.2.1.4   Image Filtering

Image filtering works on highlighting features of an image or removing certain features from the image. Image filtering is a neighbourhood operation. Filtering involves sharpening, smoothing and edge enhancement. A filter mask is created in an image to perform pixel operations inside the mask.

i. Spatial Domain Approach

Here, the spatial information or the neighbourhood information of the pixel is used to convolve the filter and get the filter and image response. The filter is moved from pixel to pixel in an image to get the corelation of mask coefficients and image intensity value over a neighbourhood. This response is calculated to define the relationship. The response provides the filtered image.

a. Linear Spatial Filtering: A linear filtering technique places the filter over image pixels. The pixel intensities are multiplied with the mask coefficients and finally added up to give the new pixel intensity values. Linear filtering can be achieved in a spatial domain. Examples of linear filtering in a spatial domain are correlation (defines correlation between pixels to find similarity between parts of image) and convolution (here, the mask is convolved both vertically and horizontally).

b. Nonlinear Spatial Filtering: Nonlinear filtering does not get a coefficient multiplied sum of intensity pixels. Here, maximum function is used instead of summation to give the highest coefficient and intensity value. Order statistics filter is a nonlinear spatial filtering, where the response is based on order or ranking of the pixels. An example of an order statistics filter is a median filter.

c. Smoothing Spatial Filtering: Smoothing spatial filtering focuses on de-blurring the blurry or foggy image and noise reduction. Noise reduction can be achieved by linear as well as nonlinear filtering. Smoothing linear filters includes averaging filter and weighted average.

d. Edge sharpening: Sharpens the edges and define crisp boundaries for the edges. It is also known as edge crisping or edge enhancement. A blurred version of image is subtracted from the original image to get an enhanced image.

ii. Frequency Domain Approach

Edges and sharp transitions represent high-frequency information of Fourier Transform, whereas low-frequency information is given by a general appearance of image over smooth area. Initially, the image is transformed into a frequency domain using Fourier Transform, and then the filtering operation is applied. On accommodating the changes, inverse Fourier Transform is applied to convert back the image into original format.

a. Smoothing Frequency Domain Filtering

A specific range of high-frequency component is attenuated for smoothing. The filtering operations include blurring, sharpening of an image, etc. Filters used for these purposes are as follows:

• Low-Pass Filters: A low-pass filter passes the signals below a certain cutoff frequency that remains un-attenuated and for signals above the cutoff frequency, the filter attenuates the signals.

  • Ideal Low-Pass Filter: Ideal low-pass filter works for sharp filtering. Here, the transition from pass band to stop band is very sharp. Filter as an image is considered where D_o represents the radii of the circle. If the radii are reduced, images become more blurred because only low-frequency information is considered, which happens to be concentrated towards the center of the circle and as we move farther to the center of circle, more high-frequency information is in consideration. So less radii takes only less frequency information into consideration and thus, completely attenuates high-frequency information outside the circle. An ideal low-pass filter H(u, v) is defined as

$$H(u, v) = \begin{array}{l} 1, D(u, v) \le D_o \\ 0, D(u, v) > D_o \end{array}$$

where D_o is the cutoff frequency and D(u, v) is the distance between point (u, v) in the frequency domain and the center of the frequency rectangle.

• Butterworth Low-Pass Filter: A butterworth low-pass filter works on the parameter filter order. The filter works on a smooth transfer function. Sharp discontinuities remain absent in a butterworth low-pass filter. Hence, no specific cutoff frequency is defined. A butterworth low-pass filter provides transition between two extremes, i.e., for high-order values, it works as an ideal filter and for lower-order values, it works as a Gaussian filter. The transfer function H(u, v) for the butterworth low-pass filter is given by

$$H(u, v) = 1/\left(1 + \left(\frac{D(u, v)}{D_0}\right)^{2n}\right)$$

Corresponding to frequency domain, the response in a spatial domain is given by the sinc function. The sinc function is inversely proportional to the radii of H(u, v). The larger the D_0 is, the more spatial sinc approaches towards an impulse. With an increase in D_0, ringing artefacts decreases which are represented by small lobes in the sinc function. Big lobe is the central lobe of the sinc function that is responsible for blurring. So, more the D_0, less the blurring there is, and the ringing effect is also small.

- Gaussian Low-Pass Filter: A Gaussian low-pass filter works for smooth filtering, which means a Gaussian filter works on a smooth transfer function and the transition between the pass band and stop band is smooth. The impulse response is smooth, resulting in no ringing. But along with less ringing, smoothing is also less as compared to the Butterworth low-pass filter. The difference between two Gaussians gives the band pass filter. Gaussian low-pass filter H(u, v) is given by

$$H(u, v) = e^{-\frac{D^{2(u, v)}}{2D_0}}$$

- High-Pass Filters: A high-pass filter passes the signals above a certain cutoff frequency that remains un-attenuated and for signals below the cutoff frequency, the filter attenuates the signals. A high-pass filter can be obtained by subtracting the low-pass filter from 1.

  - Butterworth High-Pass Filter:

$$|H(u, v)|^2 = 1/\left(1 + \left(\frac{D_0}{D(u, v)}\right)^{2n}\right)$$

  - Gaussian High-Pass Filter:

$$H(u, v) = 1 - e^{-\frac{D^{2(u,v)}}{2D_0^2}}$$

- Ideal High-Pass Filter:

$$H(u, v) = \begin{array}{l} 1, \quad D(u, v) > D_o \\ 0, \quad D(u, v) \leq D_o \end{array}$$

3.2.2 BACKGROUND SUBTRACTION

Background subtraction is the process of eliminating or subtracting background objects from the video frames.

3.2.2.1 First Difference Method

Here the first frame of video is considered to be a background frame or reference frame, and further frames are compared with this background frame and the difference between the two frames gives the foreground region; hence, the common region is the background regions that have not been changed in the frames so far. The difference between current frame and reference frame gives the foreground regions and, thus, subtracting background from the frame. This approach has good results but has a problem that it considers every object in the first frame as background; if there is any moving object in the first frame itself, the system fails to detect the background and foreground regions in the first frame.

3.2.2.2 Gaussian Mixture Method

A Gaussian mixture method [4] uses three Gaussians that store the intensity of each pixel. The first Gaussian holds pertinent pixels, or pixels showing no change in intensity. The second Gaussian holds pixels with relative changes. The third Gaussian holds pixels with sudden changes or their intensities change rapidly. The first Gaussian represent a set of pixels representing background region, and the third Gaussian represents dynamic regions. The second Gaussian holds pixels that achieve a threshold from the third Gaussian. The second Gaussian holds pixels that form a possible candidate for the static foreground region. The performance lacks compared to the frame difference method.

3.2.2.3 Collaborative Mask

Analysing the disadvantages and benefits of the frame difference method and the Gaussian method, this approach resolves the disadvantages of both. The approach creates two separate masks from the first frame as a background (mask 1) and Gaussian method (mask 2). These two masks are collaborated and a bitwise AND operation is performed. The foreground regions are the ones that both the masks have determined and are shown positive from both methods. Better object shape and accuracy is given by the first mask and checking for objects is performed by the Gaussian method. Hence, the collaborative mask [5] performs more effectively (Figure 3.6).

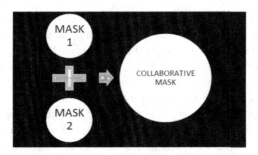

FIGURE 3.6   Collaborative Mask.

3.2.3   IMAGE SEGMENTATION

Image segmentation is the process of dividing the image into connected homogeneous regions. The homogeneity, or the basis on which these components are similar, may vary as per the need of the problem. Homogeneity can be defined on the basis of:

- Colour
- Pixel intensities
- Shape
- Motion
- Texture
- Gray value

Image segmentation can be achieved through various approaches.

3.2.3.1   Thresholding

Thresholding involves verifying each pixel grayscale value against threshold/s. The pixels which qualify or fall under the range of specified threshold values are segmented and considered as one unit. Within a given image, a number of thresholds can be defined to get different regions, where the pixels in one region are similar in terms of threshold range and are connected. By applying thresholding on a histogram segments the regions, but does not give the location information of the pixels; also, non-uniform illumination makes it difficult to segment the image.

3.2.3.2   Region Based

A region-based approach uses the region information or neighbourhood information to extract the homogeneity validations. There are techniques that work based on the region growing approach.

a. Region Growing: Region growing works on the principle of growing the region when similarity or homogeneity is encountered [1]. The seed point is selected, which represents the region, and around this point the neighbourhood is examined, and similar pixels in terms or brightness or color are added to the region and the region is expanded. Multiple seed points can be selected, each representing one region and the same approach can be applied to segment different regions. There may be a case of collision between two regions, where a pixel can fall under both the regions. In such case, the pixel is considered to be the boundary or the pixel separating two regions. This process is repeated until no particular region can be expanded further, or the whole image has been traversed.

b. Split and Merge: Region splitting works on the principle opposite to that of region growing. Region growing resorts to a bottom-up approach, i.e., region starts from a seed point and expands itself, whereas, region splitting is a top-down approach. Split and merge involve splitting and merging later. Initially a complete image is assumed to be a homogeneous region. If the assumption does not hold true, the image is split into four sub-images. Now, each sub-image is considered to be a homogeneous region and splitting continues until the whole image is homogeneous. Once splitting is complete, and the image is divided into homogeneous regions, merging is done. Two adjacent regions are considered and their homogeneity is checked; if they are similar or homogeneous in nature, considered two regions are merged to form one homogeneous region. Merging is repeated until there are no new homogeneous regions in the image.

3.2.3.3   Edge Based

Active contour is an edge-based technique. A contour (snake) is considered around the object in the image. This contour represents energy in the form of internal energy (continuity and curve bending energy), external energy (gradient) and constraint energy (external constraints). When the energy associated with the contour is a minimum, the contour touches the boundary of the object. The aim is to minimize the energy associated with the contour. Therefore, an object detection problem is reduced to an energy minimization problem.

3.2.3.4   Topology Based

A watershed algorithm is a topology-based segmentation method. The image is considered and transformed into a topographic surface. The image includes catchment basins or watershed (the cavity or basin, the slope of basin is considered watershed), minima (local minima of each catchment basin) and watershed lines (dam or lines separating the catchment basins). Segmentation is achieved by selecting connected components/catchment basins in the image. The objective is to find watershed lines. The water is hypothetically filled from each minima in all the basins; whenever the water from two basins merges, a dam is constructed at the intersection point. This dam represent the watershed line.

3.2.3.5   K-means Clustering

K-means clustering resorts to clustering the data points into various clusters. K represents the number of clusters to be achieved. Initially, data points are partitioned into k clusters randomly. For each cluster, a centroid point is calculated. For each point in the image, its distance is calculated to each cluster. The cluster with the least distance to the point is considered to be the cluster in which a point is to be included. After adding a new point, a cluster is re-formed and its centroid is computed again. This process is repeated until no point is left in the image or no cluster centroid is changed.

3.2.4 OBJECT DETECTION

Object detection refers to detecting the presence of an object of interest. Various applications require selecting a certain object and their later outputs depend on successful detection of required objects. Differentiating objects from each other and detecting a selective object of interest is known as object detection. There are various algorithms that allow efficient object detection.

3.2.4.1 R-CNN

Region-CNN (R-CNN) is a CNN (convolutional neural network)–based deep learning algorithm. Firstly, selective search is used to create region proposals of approximately 2,000 or a bounding box to classify image. Each proposal image classification is performed via CNN. All the proposals are then refined with the help of regression.

This cannot be incorporated in a real-time scenario because of high testing time for an image (approximately 47 seconds for one test image).

3.2.4.2 Fast R-CNN

Fast R-CNN depicts the functionality of R-CNN, but we give input image to CNN in contrast to proposals in R-CNN. A convolutional feature map is created, which is used to detect region proposals with the help of a ROI pooling layer which in turn is then forwarded for classification phase. It is faster than R-CNN, since, 2,000 proposals are not generated.

3.2.4.3 Faster R-CNN

Here, convolutional layers are used to generate a feature map. The feature map is then forwarded to a region proposal network that generates proposals. The proposals are then passed onto a ROI pooling layer that reshapes the proposal and classifies the proposal.

By default, there are nine anchors (box) at a position. A region proposal network reduces the number of proposals significantly; hence, less proposals are to be processed.

3.2.4.4 YOLO

You only look once (YOLO) [6] looks at parts of an image that have some confidence or probability or have an object instead of looking at proposals as applied by other algorithms. One single convolutional network makes predictions for bounding boxes and their classification is based on a probability score of similarity.

The model cannot detect small objects, but can be incorporated in real time due to its faster probabilistic parts-based classification.

YOLO is based on CNN (convolutional neural network), where the image is mapped with filters that in turn give a feature's information. It has several model variants:

- YOLO
- Tiny YOLO
- YOLOv2
- YOLOv3

3.2.4.5   SSD

A single shot multi box detector (SSD) uses one forward pass in the network to localize and classify the objects. A multi box represents the bounding box regression. It has various components and sub-components:

- Multi box
- Priors
- Fixed priors

The swiftness and precision are perfectly balanced in SSD. It also uses an anchor box and predicts bounding boxes after the convolutional layer.

3.2.4.6   R-FCN

Unlike all regional proposal network approaches, R-FCN removes a fully connected layer after ROI pooling. All the score maps are created before ROI pooling, i.e., all the complex work is done before pooling.

Average voting is performed on the basis of score maps. There is no time taking process left after ROI pooling, so it makes R-FCN even faster than faster R-CNN.

3.2.5   FEATURE EXTRACTION

Feature refers to salient and distinguishable properties of the image. Working with th whole image is a computationally costly process. Thus, selecting features from an image and performing operations to these features saves a lot of time and computational resources. The process of deriving and acquiring features from an image is known as feature extraction. Feature extraction is not only helpful for resource efficiency, but also, localising a point or region of interest, and thus, giving local control over objects/regions/features for better processing. Features may include color, texture, edges, shape, boundaries of object, etc. Feature extraction is achieved through digital differential operation (Figure 3.7).

3.3   VISION IN SINGLE AND MULTIPLE IMAGES

3.3.1   STEREOPSIS

Stereopsis or stereo vision refers to vision accomplished with the help of two or more cameras, i.e., different views from different cameras, and is considered to generate a

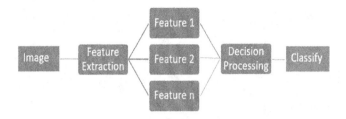

FIGURE 3.7   Feature Extraction.

three-dimensional representation of the object in the image. Depth information, which is absent in an image, can be recovered using multiple images from different cameras. Thus, stereo vision solves the ill-posed problem of computer vision [1]. Binocular stereo vision provides a way of getting depth (z-dimension) information about an object/scene from a set of two-dimensional views of the object/scene from multiple cameras. Binocular stereo vision is used by humans and animals, where each eye represents a camera and corresponding to each eye, a view of the scene/surrounding/ object is created. The views generated from both eyes are used to calculate the depth information of the scene/object. This binocular stereo vision is programmed into machines to replicate the vision of humans and animals. The goal of stereo analysis is to generate 3D scene coordinates from 2D image coordinates. Stereo analysis requires images from at least two views. To obtain depth information, the points in the left and right images (from different views) are matched. Disparity is determined from all the pixels of two images to generate a disparity map that gives the depth information. Stereopsis involves finding corresponding elements in two images and establishing 3D coordinates from 2D correspondences found during point (element) matching. The matching problem can be undertaken either by a pixel-based approach (pixel intensity values are matched) or feature-based approach (important features are matched). Pixel or intensity-based approach produces a dense disparity map, and is sensitive to illumination change, whereas a feature-based approach produces a sparse disparity map and is relatively insensitive to illumination changes (Figure 3.8).

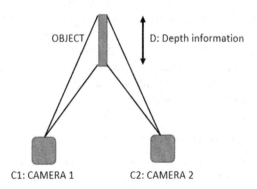

FIGURE 3.8   Stereo Vision.

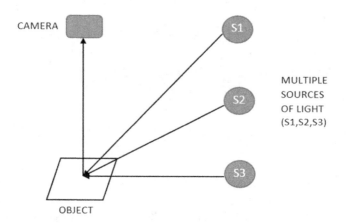

FIGURE 3.9   Photometric Stereo.

3.3.2   PHOTOMETRIC STEREO

Photometric stereo is a technique to identify a 3D shape of an object in an image with the help of different images of different brightness and different illumination conditions. Several pictures of the same object under the same viewpoint but different lightning are considered. Photometric stereo is different from binocular stereo vision by the fact that images used in binocular stereo vision are from different views, unlike photometric stereo where the views are same but different lightning conditions areconsidered (Figure 3.9).

3.3.3   SHAPE FROM SHADING

Shading represents variable levels of darkness that give a cue of the actual 3D shape. There is a relation between intensity and shape. Scene geometry is reconstructed from a given grayscale image along with albedo (surface reflectance) and light source direction. We can model the surface using surface normal for every point. Shading or the darkness levels can be used to provide depth information of a 2D image. Dark shading level corresponds to more depth, whereas less dark or light shading level corresponds to less depth. A shape from shading is generally computed when stereo vision or multiple images are not available, i.e., single image is available. An assumption is considered regarding lighting conditions known as a-priori. A shape from shading is an ill-posed problem, since many shapes can give rise to the same image. A shape from shading may also be accomplished using photometric stereo for removal of shadows and better shape reconstruction.

3.3.4   STRUCTURE FROM MOTION

An object in motion can be traced through consecutive image frames. An image has only two-dimensional information available, so structure from motion determines the structure of an object in motion from its motion information. Structure from motion uses the principle of motion parallax, which ascertains that the plane of the moving

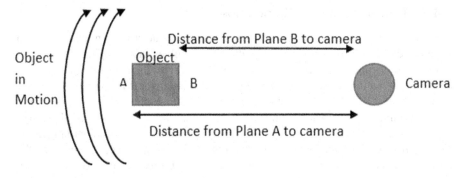

FIGURE 3.10 Motion Parallax.

object that is near the camera appears to move faster. Objects moving at a constant speed across the frame will appear to move a greater amount if they are closer to an observer or camera than they would if they were at a greater distance (Figure 3.10).

3.4 APPLICATIONS OF COMPUTER VISION

3.4.1 PERSONAL PHOTO

The wide range of camera availability has opened opportunities for applying computer vision techniques on images for better visualization and improved image quality. Personal photos can be enhanced with vision techniques; this may involve applying a filter, image smoothening, de-blurring of an image, cropping the image or feature enhancement. Applying mid-level vision and high-level vision techniques have added a greater list of functionalities associated with personal photos. Creation of a photo album of one person selectively involves vision techniques; also, features like creating a short video story of similar photos are achieved with the help of computer vision.

3.4.2 SURVEILLANCE AND SECURITY

In today's world, security at public places is a main concern for authorities. Automated video surveillance and security systems are the answer to such situations where the designed system checks for security are based on certain defined conditions. To achieve high efficiency and reduce human monitoring, an automated system serves the task that no longer requires human monitoring to accomplish security. Computer vision can target certain scenarios that jeopardize the security or safety; framework EDCAR [7] works on automatic video surveillance, for activities such as stealing, unattended object, crowd activity monitoring, tracking a person/object, fighting, etc. Video surveillance can also help fire scenarios and alert authorities. Thus, computer vision techniques enhance and provide efficient and effective frameworks for smart video surveillance systems.

3.4.3 Optical Character Recognition (OCR)

Optical character recognition converts the textual or character data in the image into editable form. Optical character recognition [8] involves block segmentation (portioning homogeneous connected components/blocks), line segmentation (segmenting each line from a block), word segmentation (segmenting words from lines), character segmentation [9] (segmenting characters from words) and character recognition (understanding the character by matching it to the stored character database and identifying the character with its closest similarity to the syntax of given language). The segmented characters are matched with the syntax of language to recognize the character. Optical character recognition is useful in preserving the content of old, significantly important documents with degraded script. Optical character recognition works on the segmentation and pre-processing feature of computer vision.

3.4.4 Face Recognition

Face recognition offers a range of applications discussed in article [10]. Extracting facial features from the image and then matching these features to the features stored in a database provides authentication approval. Only the authorized person can have access rights to the application/system. Face recognition is used in identity matching, to unlock mobile phones/digital devices, smart attendance system etc.

3.4.5 Virtual Reality

Immersive multimedia has been popular in recent times. Virtual reality provides for a near real-world experience in multimedia applications. Virtual reality gives the essence of immersion in the scene. The user finds themself immersed in the application or finds themself to be a part of the application. Virtual reality movies provide a near-real immersive movie experience. Virtual reality can be used in fire safety training [11], and various simulation applications like a swimming training model [12], and an airplane simulator uses the concept of virtual reality. Virtual gaming also has proven to be an important application of computer vision.

3.4.6 Natural Language Processing (NLP)

Natural language processing concerns understanding and recognising human written language. NLP is comparatively complex, provided that human handwriting can be complex and difficult to read and understand. Writing styles and stroke understanding involve complex computational vision processing. Natural language processing has its application in the field of banking, authentication by human handwriting, recognition of signature, etc.

3.4.7 Assistive Technologies

Computer vision has found a greater range of application in assistive technologies for persons with disabilities [13]. Visually impaired people find a lot of difficulties in routine work that require vision. Understanding the environment and providing a description about the environment and navigating these people with the help of vision provides a lot of assistance and can work as eyes for the visually impaired. Understanding the scenario from visual images involves recognising the types of objects that may be an occlusion in the path of walking for a person, and also distinguishing between static (stable) objects and dynamic (moving) objects. Assistive technologies have been an area where research is extensively continued, provided that a large amount of people suffer from visual impairment and providing assistance through vision technology turns out to be a great application.

3.4.8 Biometrics and Forensics

Identification and authentication are fundamentals for any system to authorize only allowed persons to have access and work on an application/system. Validating the identity of a person and authorizing the level of access which has been set for that user remains an important task at hand. For this purpose, biometrics provide a unique distinguishing feature for a person. These features are stored in a database and whenever there is an access request, the biometrics are matched with the stored features and the similarity score above a threshold validates the identity of user and provide access. Biometrics may include fingerprints, retina features, etc. In recent research, elbow prints and knee prints have proved to be good biometrics to uniquely identify a person. The field of forensics [14] involves studying and understanding forensic data ranging from fingerprints, blood, stains, etc. Understanding the features contained in this data and providing evident results is accomplished with the help of computer vision.

3.4.9 Astronomy and Cosmic Study

The study of celestial bodies and cosmic phenomena require computational and complex work on satellite image and spectral images. Understanding satellite images and providing concrete data evidence is supported by computer vision. These bodies emit radiations and comprise of different elemental structures that emit heat and other energy. Understanding these wavelengths, frequencies and energy functions through spectral imaging gives computer vision a lot of exploration opportunities and so far there has been sufficient support from computer vision.

3.4.10 Machine Inspection

Industrial machinery require regular maintenance, provided that the inspection is done regularly and effectively. Computer vision can be used to identify problems or inspect machines using images of the machine as proposed in [15]. The inspection may include identifying an uneven structure, rusting on parts of a machine, breakage or cracks in the segments. Uneven breaks in texture and shapes or structure of a machine gives the insight of faulty machinery. Also, spectral images can also be used to detect any leakage from pipes and hoses. Thermal images can be used to monitor the heat emission from pipes, etc. and can further be used to identify which emit more heat; thus, irregularities in the structure can be determined. These applications help to identify these maintenance points and perform the machine inspection with less amount of time and provide better results.

3.4.11 GAMES

In recent years, the gaming industry has evolved. Console screen–based games are now being switched over to immersive virtual reality gaming. A basketball game using computer vision is proposed by [16]. These games involve virtual reality headsets; also, devices to capture the body movements and provide an immersive gaming experience that involve visual, sound and movement representations. Holographics are also evolving to convolve holographic views with games and provide a more interesting, immersive and real gaming experience. These holographic views and virtual reality gaming are the applications where computer vision is gaining popularity.

3.4.12 PANORAMA STITCHING

A panorama is a merged image consisting of various consecutive images of the same environment, where each image shifts in one direction. The final merged images formed by all these consecutive images, or by stitching all these images to form one single image with wider field of view, is known as a panorama. Generally, the aim of panorama stitching is to give a 360-degree view of the environment through an image. Panorama are commonly available in mobile or camera devices. Computer vision techniques capture the common features from two consecutive images or point of interest and finally match them and form a new union image on the basis of these common interest points.

3.4.13 ROBOTICS

Advancements in automation and computer vision techniques have provided a foundation for robotics or smart robots. Providing these robots with vision capabilities, providing them a description of the environment and making decisions for the robots to react. Assisting robots in finding the way/path through camera images and thus making the robot understand the environment requires significant research efforts. Robotics require information about the immediate real world around the robot and hence becomes very important to capture, understand and react to the same.

3.4.14 IMAGE FUSION

Image fusion refers to merging two or more images together. The fusion might involve applying texture from one image to another, changing the background or foreground of an image by reference to any other image. Image modification and creating custom images based on user preference is handled by computer vision through image fusion. Image fusion has applications in the field of image processing, product development, forensics, sketch formation, applying skin texture to a sketch to predict the person, etc.

3.4.15 NAVIGATION

Intelligent navigation systems learn from the environment through cameras and provide a better navigation to the user based on the path details and destination information. Navigation using computer vision is proposed by [17]. The capturing of immediate surroundings is captured and these are used to find the best possible path for the user. Smart navigation in cars provide a field of view through a parking camera, etc. and thus help in better navigation solutions.

3.4.16 3D RECONSTRUCTION

Reconstructing the 3D scene environment from a 2D image is called 3D reconstruction. The depth information or z-dimension is lost in the image. Applying computer vision techniques can retrieve this depth information. For this calculation, shape from shading, stereo vision, shape from motion, photometric stereo and shape from texture techniques can be used.

3.4.17 TRAFFIC MANAGEMENT AND MONITORING

Traffic signals based on static information remain fixed, i.e., the traffic light time mechanism remains unchanged irrespective of the number of vehicles. Thus, computer vision–enabled traffic signals may be employed to work on real-time data and manage the signals accordingly. Also, monitoring traffic rule violations is handy with vision techniques. Cameras installed can track the vehicle registration number that committed the traffic violation and can be monitored accordingly.

3.4.18 DEFENCE AND MILITARY APPLICATIONS

Defence establishments need the latest advancements in technologies to maintain the security parameters. Missile launching systems require accurate pin point data and information; computer vision can provide these useful contributions by the help of satellite images. Also, thermal images provide useful vision in nighttime for patrolling services and other operations. Understanding the contours out of thermal images is solved by computer vision. Providing better navigation and assisting UAVs (unmanned aerial vehicles) with vision [18] are the important applications of computer vision. A smart border monitoring system can be employed to detect intrusion activities.

3.4.19 TEMPLATE MATCHING

Template matching aims to match two images and find similar object or environment in the given images. Identifying the features of images and finding any similarities with those features in the target image are achieved through computer vision.

3.4.20   OBJECT RECOGNITION

Object recognition involves detecting and identifying the object and hence recognising it from a certain group of objects. Images may contain many objects and recognising these objects on the basis of their features. Objects may not be clear, occlusion might hinder the view of objects and also, low illumination affects the recognising performance. These problems can be handled well by pre-processing and then apply vision techniques to recognize the objects.

3.4.21   MOTION TRACKING

Tracking the motion of an object/person works on consecutive frames where an object is tracked based on certain features that may be fetched in next frame through template matching. Feature tracking may include fuzzy logic, and features must be uniquely identifiable for the system to track the object accurately. Tracking the trajectory of the motion followed by the object of interest requires various computational vision procedures.

3.4.22   CONTENT-BASED IMAGE RETRIEVAL

Image retrieval techniques have advanced much over recent years. Content-based image retrieval involves image retrieval techniques based on user expectations or query-based image retrieval. The query is converted to represent some features, and these features are then looked for in the target image, based on similarity score of query features and image features; the image retrieved has the closest similarity to the query. Such an image retrieval technique is called content-based image retrieval. Computer vision is the idea behind content-based image retrieval.

3.4.23   WEATHER FORECASTING

Weather forecasting involves understanding the wind currents and cloud movement patterns to predict or forecast the weather. These movements can be predicted by analysing satellite images over a certain time that include cloud distribution, temperature and humidity information, wind speeds and directions. Computer vision can help facilitate in flood management [19]. Thermal and satellite images prove to be important factors to be understood and predict the weather by passing this image understanding to weather forecasting system models.

Along with the applications discussed, computer vision is turning out to be the elementary basis for any activity involving understanding images and making predictions or decisions based on gained insight or understanding of the image. The applications of computer vision go as far as image data for that application is available, and it requires any sort of understanding from the image.

3.5 PROBLEM IMPLEMENTATION: UNATTENDED OBJECT DETECTION

Automated surveillance systems are the need of today, providing an intelligent and smart monitoring system that that keeps an eye on the crowd activity in public areas. From security perspectives, any activity that requires urgent attention needs to be identified early. A real-time system for unattended object detection provides this feedback, monitors the crowd through a wide coverage of surveillance cameras and alarms/alerts authorities whenever any instance of an unattended object is encountered. Unattended objects left in public places can pose serious risk to the security and safety. Hence, automatic detection of unattended object event analyses the video frame by frame [20] and performs certain condition check before alarming authorities of the event (Figure 3.11).

3.5.1 What Is an Unattended Object?

An unattended object is termed as any object that is introduced to a place along with its owner, and after some time, the owner of the object is not around the neighbourhood of the object (defined by d meters) and the owner is not in the neighbourhood of the object for time equal to or more than the specified time for an object to be termed unattended (say t seconds); also, if the owner has some social relations, these objects can be attended by people of their social context and will not be termed unattended. In the case of multiple owners, if any one of the owners is a nearby object, the object shall be attended for that time and distance.

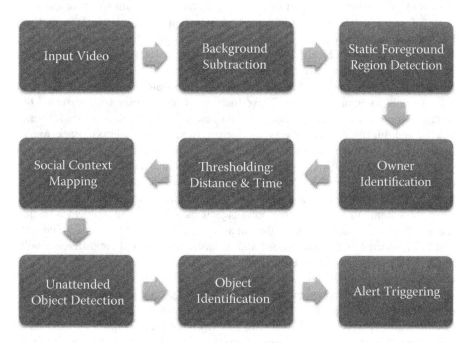

FIGURE 3.11 Approach for Unattended Object Detection.

3.5.2 INPUT VIDEO

The input video is generally a live surveillance camera feed. Based on the char-acteristics of the camera, video is defined by fps (frames captured per second). These frames are actual image sequences that form the complete video when run in continuation. To account for more scenarios in the system, a 360-degree-view camera can be considered. Also, multiple cameras are to be used to track the owner on a much wider scale. The live feed of a surveillance camera or the recorded video may also be used to provide data for the system top work upon. This video work is adata set of the system. Also, different scenarios with a mixed environment can be used to check the robustness and accuracy of the system.

3.5.3 BACKGROUND SUBTRACTION

The input video is transmitted to the system to start the working procedure. The first step remains to identify objects as background objects or the foreground objects. There are objects in the frame or video that are a part of the background for which the system need not process, since these background objects will remain static or fixed. These objects do not require system attention and hence to reduce the computing or filter the objects that are of least interest to the application in hand, background subtraction is performed. Background subtraction refers to eliminating all the back-ground objects present in the image. All that is left after filtering or subtracting background objects from the image is considered the foreground. So, for a given image sequence, the objects present in the image or the scene contain either back-ground objects (which are a part of background and shall remain fixed for the life-cycle of event detection) or foreground objects (objects that are not a part of background, or represent a newly added object to the image that was not in previous frames). Out of these objects, foreground objects are objects of interest to process further. So, background subtraction takes surveillance feed/video as input, subtracts background objects and outputs the background eliminated or filtered image that contains only foreground objects. These foreground objects may be represented by contours or connected regions displayed in a binary image to filter out background objects and highlight only objects of interest (foreground objects). Background subtraction can be performed using frame difference method, Gaussian mixture model [21] or collaborative approach. In a frame difference method, the initial frame of the video is considered to be the background frame and consecutive frames are then compared with the background frame. The background frame is subtracted from the current frame to give the foreground objects as a result. The problem with the frame difference method is that, of the initial frame consist of any dynamic object, it will be considered as the background and in further frames, and the difference will result in false positives and false negatives. The Gaussian mixture method [4] is based on an adaptive background mixture model [21] and uses three Gaussians to represent a pixel. One Gaussian represents background/pertinent pixels, which remain fixed and do not show any change in intensity values (part of background object); the second Gaussian stores pixels with relatively low movement and remains static or fixed after some time. The third Gaussian stores a pixel with varying intensities or dynamic pixels that represent a dynamic foreground or objects in motion. The Gaussian method performance is hampered from sudden illumination changes and low lighting conditions. A collaborative method [5] combines both the frame difference method and Gaussian mixture method. The result from both the frame and Gaussian method is mapped to perform bitwise and operation, i.e., both results are operated by logical AND operation on pixels. The operation results in foreground objects with less false positives and consider an object to be the foreground only if it is termed as foreground by both these methods.

3.5.4 STATIC FOREGROUND REGION DETECTION

Background subtraction gives the result of all the foreground objects available in the current frame. The next step is to refine these objects and filter out less useful objects to be forwarded for processing. Foreground objects can be static or dynamic. Static foreground objects are the objects that are not part of the background, and are introduced in the frame through some motion, but after a certain time period these objects get static or do not move at all. These objects are termed a static foreground. Dynamic foreground objects represent the objects that are not a part of the background, and hence are introduced in the current frame and are still in motion, i.e., their position is not fixed/stable/static. For the given problem statement, only static foreground objects can be a possible candidate of an unattended object, so a static foreground region is selected, which contains only static foreground objects and dynamic foreground objects are filtered out. Static foreground region detection and background subtraction work together. So, using background subtraction techniques, not only background objects are subtracted but also dynamic foreground objects and the resultant frame consists only of a static foreground region containing static foreground objects.

3.5.5 OWNER IDENTIFICATION

Static foreground region detection gives the possible candidates of an unattended object subject to time rule, distance rule and ownership. For a static foreground object to be termed unattended, the owner of the object should first be identified. Liao et al. [22] implement selective tracking on the owner. Once the owner is identified, the current frame is checked for the neighbourhood of the object against its owner. If the owner is not in the range of the neighbourhood of the object (say d meters), the distance rule is achieved. Also, if the owner is not found in the neighbourhood of the object for more than t seconds, then the time rule is achieved. After owner identification, if the owner satisfies both the time rule and distance rule (the owner is not present in the neighbourhood of object defined by distance threshold (d meters) for more than time threshold (t seconds)), the object is termed unattended.

The next step involves identification of the owner to check for thresholding conditions. The owner can be achieved, only with a relationship between object and owner that can be defined from the first frame in which both the owner and object entered the scene as dynamic foreground objects. So, backtracking is performed; the current frame is traced back to few frames. The back trajectory followed by an object is used to get the location of an object in the previous frame. Template matching can be used for this, along with fuzzy logic, where the trajectory is determined with the most probable path given by logic. The frames are traced back to the point when the contour (connected component in binary image) of the object is the union of object contour as well as contour of owner; in other terms, the owner is holding the object and there is a contact between the owner and object that can be represented as a union of human contour and object contour. From this frame, the object contour is subtracted from the union contour to give the owner contour. Human contour tracking can also be performed by matching the similar contour shapes in consecutive frames to track the owner contour, but this technique may be subject to vulnerabilities, because from a contour track two or more contour trajectories can be detected; also, two contours might come close enough to form a large union contour, so this technique may hamper the owner tracking in certain conditions. Other criteria would be to identify the owner on the basis of unique features of the owner. From the backtracked frame, features can be extracted from the owner. These features may include facial features, silhouette [23], appearance model [24], human skin color information [25], human activity [26], clothes and style of walking [27]. Any of the features extracted can be represented by a feature vector. This feature vector is searched for in the next frames; the object or person that has the closest similarity of feature with the feature vector defined is considered to be the owner or the same person whose features were extracted. This procedure is followed to identify the owner of the object.

3.5.6 THRESHOLDING

PETS2006 [28] presents two threshold conditions of time and distance:

i. **Distance Rule: The distance rule specifies that the owner of the object should not be present in the neighbourhood of the object for object to be termed as unattended. The neighbourhood is considered as a circle of radius d meters around the object, where the owner presence is to be checked. If the owner is not present in this neighbourhood circle, the distance rule is satisfied. A graphical circle bound is drawn around the object of radius d meters to check for the owner presence.**

ii. **Time Rule: The time rule specifies that the owner of the object should not present in the neighbourhood of the object for more than t seconds. The neighbourhood is defined by the distance rule. If the distance rule holds true for t seconds, then the time rule holds true as well. This means that the owner of the object is not present in the neighbourhood of object (d meters) for more than t seconds time period. If this condition is true, the time rule is satisfied.**

If both the distance rule and time rule are satisfied, then the object is termed asan unattended object. The owner of the object is not around the neighbourhood of the object and the owner is not in the neighbourhood of the object for more than the specified time.

3.5.7 Social Context Mapping

Social context relation is an important factor in determining the ownership of an object. Considering an example, a family coming together and the person holding a bag, placed the object and left, but the other family members are still in the neighbourhood of the bag, this scenario will term a bag as an unattended object. The method [29] is used in the framework to infer the social relations between the individuals in a scene. To avoid cases like these, social context relations of the owner are to be understood. Human activity recognition or social actions like talking or holding hands can be made a basis to determine social context relations. A model like the Hidden Markov Process Model can be trained on activities like these, and whenever these activities are accounted in the context of the owner, the other entities/people involved in social context relations are also considered as object owners. The object may be termed unattended only when no owner is around the neighbourhood of the object and fulfils both the distance and time rule. If any of the owner is present in the neighbourhood of the object, the object shall remain attended and no further processing is to be done.

3.5.8 Unattended Object Detection

The possible candidate of an unattended object is processed to identify the owner. Once the owner is identified, the owner is tracked to check the thresholding parameters given by distance rule (check for owner's presence in the neighbourhood of the object) and time rule (checks for distance rule happening for the time threshold), the social context relations are identified and checked. If all the owners (given by social context relations) satisfy the distance rule and time rule, the object is termed an unattended object. Hence, an unattended object is detected for the given scenario.

Algorithm: Unattended Object Detection

A. If object contour belongs to static foreground
   Then,
   1. Go back f frames
   2. Find union contour containing object contour
   3. Set owner contour = Union Contour − object contour
   4. Extract Features of owner represented by owner contour
   5. Get owner location details
   6. Check owner.location > d meters & time > t seconds
   7. Repeat above steps 4–6 for all owners
   8. If for all owners, owner.location > d meters & time > t seconds
   9. Set object = unattended baggage

3.5.9 OBJECT IDENTIFICATION

Once the unattended object is detected, the object identification is performed to check the type of object and predict the security concerns of the object. In the example discussed, if no owner (family members) is present in the neighbourhood of a bag for a time greater than the time threshold, the object is termed unattended. After this, the object is directed to the object identification phase. The object identified can be done by a number of models, generally a CNN (convolutional neural network) is used to identify the object. The YOLO model proposed by Redmon et al. [6] is used for object identification. The object identification phase will result in an object type bag for the above discussed example.

3.5.10 ALERT TRIGGERING

On successful detection of an unattended object, an alert is to raise the alarm to authorities with the event. The alert involves the location of object, along with the type of object. The object is surrounded by a bounding box along with its co-ordinates location.

The owner can also be further tracked to help the authorities, using the same features of the owner that were extracted in the early stages. Multiple cameras can be used for better tracking of the owner, where all the cameras are connected in a distributed network fashion and the features of the owner can be sent to all the cameras, and each camera can then search for the owner using the feature vector of the owner.

3.5.11 SPECIAL CASES: OCCLUSION AND MULTIPLE EVENTS

The system may also encounter some special cases apart from ones already discussed (multiple owners and multiple cameras). These include occlusion in the view of an unattended object and multiple unattended events happening at the same time.

To handle occlusion, an occlusion time threshold (Th) can be set, proposed by Tian et al. [4]. If the object is termed static in time less than the occlusion time threshold, then continue the same procedure discussed above. Otherwise, if the object is not detected as static for a time greater than/equal to occlusion time threshold, then terminate the process and no processing is to be done.

Algorithm: Occlusion Handling

1. Detect static foreground regions
2. If not detected for time >= Th then terminate
3. Else, go back when object became static (Ts),
4. If Ts >=t, Time rule (t seconds) satisfied.

If multiple unattended objects are detected at the same time, getting owner details and managing them uniquely may be complex. To solve this problem, in the backtracking step, the union contour is assigned one unique id. The individual owner and object are bound by the object-owner id pair. For each event, this id maintains uniqueness for each owner and object pair. This can result in efficient and unique identification of objects and owners. And, multiple unattended object events can be detected simultaneously.

3.5.12 IMPLEMENTATION AND RESULT

The implementation of the system is done in the Python programming language. The tool used for programming is SPYDER (tool under the framework of ANACONDA). Various algorithms helped to form the basis of the system include the following:

- Connected component analysis
- Gaussian mixture model [4]
- YOLO [6]
- Frame difference method

Various libraries are used, such as:

- cv2 (open computer vision library): It is used to process the image, work with pixel information, and accomplish the connected component contouring using the in-build FindContours procedure.
- numpy (supports multi-dimensional array and high mathematical functions),
- imageai (object detection library): It is used for image detections in the YOLO model.

Previous works either considered the first frame as the background for background subtraction or used the GMM (Gaussian mixture model) for background subtraction. Both of these methods had their drawbacks. The first frame as the background considered every object in the first frame of the video to be a part of the background, be it a person or any other object. Consecutive comparison of frames gives false positives of a static object. The GMM method did not have a high accuracy compared to the first frame method.

The proposed background subtraction used both of these methods and created different masks from individual methods. These masks are then used to create a collaborative mask using BITWISE-AND operation. This removed the inconsistencies of the first frame method and gave better results than GMM. The occlusion scenarios, which hampered performance in previous work, is managed to give more accuracy and handle scenarios with occlusion.

The system after parsing a number of frames checks for the thresholding criteria if the conditions are true for that object. The result shows the unattended object with a bounding box around it along with alarming text and location of the object with object identification probability (Figure 3.12).

```
In [3]: runfile('C:/Users/Lenovo/.spyder-
py3/new.py', wdir='C:/Users/
Lenovo/.spyder-py3')
backpack   :   55.83977699279785   :   [131,
324, 180, 391]
-----------------------------------
backpack   :   55.83977699279785   :   [131,
324, 180, 391]
-----------------------------------
```

FIGURE 3.12 Object Identification.

Three data sets were used:

1. PETS2006 [28] containing 7 video sequences
2. OWN data set (containing 6 video sequences)
3. ABODA data set [30] (11 video sequences)

Table 3.1

The detection score consists of videos that showed true positives only. In the ABODA data set [29], some video sequences resulted in true positives and false positives. The detection score has not been considered for such cases having true positives and false positives. The detection score comprises results of those sequences that not only showed true positives but also no false positives.

Approach 1 uses the first frame as the background method for background subtraction.

Approach 2 uses the Gaussian mixture model for background subtraction.

The system performed well in different environments, removing problems identified in the previous system; hence, incorporating more application areas of the system. We created our own data set involving different types of scenarios. The data set has six video sequences at a frame rate of 29 frames per second. The Gaussian mask is prone and is less efficient if sudden lighting changes occur. Bad lightning conditions and sudden lighting changes may produce false positives and may degrade the system performance. Some scenarios from the ABODA data set [30] experienced sudden lighting changes and the system parsed false results that reduced the detection score (Figures 3.13, 3.14, and 3.15).

TABLE 3.1

Unattended Baggage Detection Results

S. NO	Dataset	DETECTION SCORE (%)		
		Approach 1	Approach 2	Approach Used
1.	PETS2006	85.71 (6/7)	71.42 (5/7)	85.71 (6/7)
2.	OWN DATASET	83.33 (5/6)	100 (6/6)	100 (6/6)
3.	ABODA	54.54 (6/11)	72.72 (8/11)	63.63 (7/11)
Total		70.83 (17/24)	79.16 (19/24)	79.16 (19/24)

FIGURE 3.13 PETS2006 Data Set Result.

FIGURE 3.14 ABODA Data Set Result.

FIGURE 3.15 OWN Data Set Result.

3.6 CONCLUSION

This chapter provides insight about the computer vision and concerned fundamentals and various applications. Computer vision provides wide vision capabilities and helps computers to see and understand the real world from given images. The data acquisition phase consists of some physics involved in imaging and radiometry. The ability to see things, understand them and provide intelligent decisions are contained in computer vision. Computer vision is combined unit containing preprocessing, selection of areas of interest, processing on selected areas and intelligent decision making. Vision in multiple images can be performed to gain depth information (z-dimension) of the image and other image information using stereopsis, photometric stereo, structure from motion, etc. A single image can also be used to get depth information using the shape from the shading model, where shading represents the variable level of darkness and, hence, determine depth information. We implemented the discussed computer vision techniques and methods to solve a real-life problem of unattended object detection. The system proposed works well in occlusion and provides better results compared to previous systems. The system involves pre-processing in input video frames and selects the owner using backtracking. The features of the owner are extracted to further track the owner. The owner is tracked for two threshold conditions. Fulfilling these conditions for all the given owners provided by social context relationships resolves ownership ambiguities. Multiple cameras are used to use multiple view planes [31] and 360-degree camera methods [32] are also supported, providing a better tracking range and increasing the field of view for a detected event. Also, the system handles multiple detection events at the same time and assign a unique object-owner id to each instance for better tracking and effectiveness. Object identification performed using CNN provides more detailed results about the type of object left unattended. Thus, the system successfully explored the techniques of computer vision and provides a better automated smart surveillance system for detecting an unattended object event.

REFERENCES

[1] M.K. Bhuyan (2019). *Computer Vision and Image Processing: Fundamentals and Applications*, CRC Press, USA, ISBN 9780815370840 – CAT# K338147.

[2] D. Forsyth & J. Ponce (2003). *Computer Vision-A Modern Approach*, Pearson Education.

[3] R. Szeliski (2020). *Computer Vision- Algorithms & Applications*, Springer.

[4] Y.-L. Tian, R. Feris & A. Hampapur (Oct 2008). "Real-Time Detection of Abandoned and Removed Objects in Complex Environments," *The Eighth International Workshop on Visual Surveillance – VS2008*, Marseille, France, inria-00325775.

[5] P. Narwal & R. Mishra (June 2020). "Unattended Baggage Detection System for Occlusive Complex Environment Using Collaborative Mask," *The International Journal of Scientific & Engineering Research (IJSER)*, Volume 11, Issue 6.

[6] J. Redmon, S. Divvala, R. Girshick & A. Farhadi (2016). "You Only Look Once: Unified, Real-Time Object Detection," *2016 IEEE Conference on Computer Vision and Pattern Recognition (CVPR)*, Las Vegas, NV, pp. 779–788.

[7] L. Caruccio, G. Polese, G. Tortora & D. Iannone (2019). "EDCAR: A Knowledge Representation Framework to Enhance Automatic Video Surveillance," *Expert Systems with Applications*, Volume 131. https://doi.org/10.1016/j.eswa.2019.04.031.

[8] R. Sharma & B. Kaushik (2020). "Offline Recognition of Handwritten Indic Scripts: A State-of-the-art Survey and Future Perspectives," *Computer Science Review*, Volume 38. https://doi.org/10.1016/j.cosrev.2020.100302.

[9] I. Pattnaik & T. Patnaik (2020). "Character Segmentation on Degraded Printed ODIA Script," *International Journal of Computer Sciences and Engineering*, Volume 8, Issue 4, pp. 43–45.

[10] A. Insaf, A. Ouahabi, A. Benzaoui & A. Taleb-Ahmed (2020). "Past, Present, and Future of Face Recognition: A Review," *Electronics*, Volume 9, p. 1188. https://doi.org/10.3390/electronics9081188.

[11] S. Morélot, A. Garrigou, J. Dedieu & B. N'Kaoua (2021). "Virtual Reality for Fire Safety Training: Influence of Immersion and Sense of Presence on Conceptual and Procedural Acquisition," *Computers & Education*, Volume 166, p. 104145. https://doi.org/10.1016/j.compedu.2021.104145.

[12] X. Zhu & F. Kou (2021). "Three-Dimensional Simulation of Swimming Training Based on Android Mobile System and Virtual Reality Technology," *Microprocessors and Microsystems*. Volume 82, p. 103908. https://doi.org/10.1016/j.micpro.2021.103908.

[13] M. Leo, G. Medioni, M. Trivedi, T. Kanade & G. Farinella (2016). "Computer Vision for Assistive Technologies," *Computer Vision and Image Understanding*, Volume 154. https://doi.org/10.1016/j.cviu.2016.09.001.

[14] M. Auberson, S. Baechler, M. Zasso, T. Genessay, L. Patiny & P. Esseiva (2016). "Development of a Systematic Computer Vision-based Method to Analyse and Compare Images of False Identity Documents for Forensic Intelligence Purposes–Part I: Acquisition, Calibration and Validation Issues," *Forensic Science International*, Volume 260, pp. 74–84, ISSN 0379-0738. https://doi.org/10.1016/j.forsciint.2016.01.016.

[15] R. Manish, A. Venkatesh & S. Ashok (2018). "Machine Vision Based Image Processing Techniques for Surface Finish and Defect Inspection in a Grinding Process," *Materials Today: Proceedings*, Volume 5, pp. 12792–12802. https://doi.org/10.1016/j.matpr.2018.02.263.

[16] M. Hu & Q. Hu (2021). "Design of Basketball Game Image Acquisition and Processing System Based on Machine Vision and Image Processor," *Microprocessors and Microsystems*, Volume 82, p. 103904, ISSN 0141-9331. https://doi.org/10.1016/j.micpro.2021.103904.

[17] S. Xie, Jun Luo, J. Rao & Z. Gong (2007). "Computer Vision-based Navigation and Predefined Track Following Control of a Small Robotic Airship, *Acta Automatica Sinica*," Volume 33, Issue 3, pp. 286–291, ISSN 1874-1029. https://doi.org/10.1360/aas-007-0286.

[18] A. Al-Kaff, D. MartÃn, F. GarcÃa, A. de la Escalera & J. M. Armingol (2018). "Survey of Computer Vision Algorithms and Applications for Unmanned Aerial Vehicles," *Expert Systems with Applications*, Volume 92, pp. 447–463, ISSN 0957-4174. https://doi.org/10.1016/j.eswa.2017.09.033.

[19] U. Iqbal, P. Perez, W. Li, & J. Barthelemy (2021). "How Computer Vision Can Facilitate Flood Management: A Systematic Review," *International Journal of Disaster Risk Reduction*, Volume 53, p. 102030, ISSN 2212-4209. https://doi.org/10.1016/j.ijdrr.2020.102030.

[20] P. Narwal & R. Mishra (Nov 2019). "Real Time System for Unattended Baggage Detection," *Proceedings of the International Research Journal of Engineering and Technology (IRJET)*, Volume 6, Issue 11, No. 242.

[21] C. Stauffer & W.E.L. Grimson (June 1999). "Adaptive Background Mixture Models for Real-Time Tracking," *Proceedings of IEEE Conference on Computer Vision and Pattern Recognition*, Volume 2, pp. 246–252.

[22] H.-H. Liao, J.-Y. Chang & L.-G. Chen (2008). "A Localized Approach to Abandoned Luggage Detection with Foreground-Mask Sampling," *IEEE Fifth International Conference on Advanced Video and Signal Based Surveillance*, Santa Fe, USA, pp. 132–139.

[23] I. Haritaoglu, R. Cutler, D. Harwood & L. S. Davis(1999). "Backpack: Detection of People Carrying Objects Using Silhouettes," *Proceedings of the Seventh IEEE International Conference on Computer Vision*, Kerkyra, Greece, pp. 102–107, Volume 1. https://doi.org/10.1109/ICCV.1999.791204.

[24] A. Senior, H. Arun, L. Ying, B. Lisa, S. Pankanti & B. Ruud (2006). "Appearance Models for Occlusion Handling," *Image and Vision Computing*. Volume 24, pp. 1233–1243. https://doi.org/10.1016/j.imavis.2005.06.007.

[25] H. Liao, J. Chang & L. Chen (2008). "A Localized Approach to Abandoned Luggage Detection with Foreground-Mask Sampling," *2008 IEEE Fifth International Conference on Advanced Video and Signal Based Surveillance*, Santa Fe, NM, USA, pp. 132–139. https://doi.org/10.1109/AVSS.2008.9.

[26] S. Lu, J. Zhang & D. D. Feng (2007). "Detecting Unattended Packages Through Human Activity Recognition and Object Association," *Pattern Recognition*, Volume 40, pp. 2173–2184. https://doi.org/10.1016/j.patcog.2006.12.013.

[27] C. BenAbdelkader & L. Davis (2002). "Detection of People Carrying Objects: A Motion-Based Recognition Approach," *Proceedings of Fifth IEEE International Conference on Automatic Face Gesture Recognition*, Washington, DC, USA, pp. 378–383. https://doi.org/10.1109/AFGR.2002.1004183

[28] PETS (2006). Benchmark Data. http://www.cvg.rdg.ac.uk/PETS2006/data.html

[29] J. Ferryman, D. Hogg, J. Sochman, A. Behera, J. Rodriguez-Serrano, S. Worgan, L. Li, V. Leung, M. Evans, P. Cornic, S. Herbin, Stefan Schlenger & M. Dose. (2013). "Robust Abandoned Object Detection Integrating Wide Area Visual Surveillance and Social Context," *Pattern Recognition Letters*, Volume 34. pp. 789–798. https://doi.org/10.1016/j.patrec.2013.01.018.

[30] K. Lin, S. Chen, C. Chen, D. Lin & Y. Hung (July 2015). "Abandoned Object Detection via Temporal Consistency Modeling and Back-Tracing Verification for Visual Surveillance," *IEEE Transactions on Information Forensics and Security*. Volume 10, no. 7, pp. 1359–1370.

[31] S. Khan & M. Shah (2009). "Tracking Multiple Occluding People by Localizing on Multiple Scene Planes," *IEEE Transactions on Pattern Analysis and Machine Intelligence*, Volume 31, pp. 505–519. https://doi.org/10.1109/TPAMI.2008.102.

[32] D.T. Singh & D. Kushwaha (2016). "Tracking Movements of Humans in a Real-Time Surveillance Scene," *Proceedings of Fifth International Conference on Soft Computing for Problem Solving*. https://doi.org/10.1007/978-981-10-0451-3_45.

4 Virtual Reality in Social Media Marketing: The New Age Potential

*Sheetal Soni[1], Kaustubhi Shukla[2],
Usha Yadav[1] and Harjeev Singh Ahluwalia[3]*
[1]National Institute of Fashion Technology, Jodhpur,
Rajasthan, India
[2]ImaginXP, Pune, Maharastra, India
[3]Value Labs LLP, Hyderabad, Telangana, India

4.1 INTRODUCTION

People's lifestyles have changed in unprecedented ways due to the worldwide COVID-19 pandemic. Many around the globe are staying indoors, at home or otherwise, and limiting all physical and social contact. Friends and families have moved to interact via Zoom calls, and those brave enough to venture outdoors organize drive-by birthday parties. Employees in the corporate ecosystem are spending innumerable hours on screens, talking to clients and colleagues [1,2]. In addition to this, consumer behavior has also changed due to the Internet, and so has the way companies go about doing their businesses. The increasing use of the Internet and social media have created challenges for marketers looking to combine conventional marketing strategies with newer ways of drawing the attention of customers. It is this effort that has led marketers to turn towards virtual reality. Virtual reality (VR) is one of the newest buzzwords in the marketing era that is now paving ways for marketers to develop new strategies to interact with customers and increase awareness towards their brands.

In 2020, consumer and enterprise VR revenue amounted to USD$3.89 billion, a reduction from the previous year's revenue, resulting from the impact of the pandemic. In 2021, VR revenues are expected to once again climb to USD$4.84 billion [3].

4.1.1 VIRTUAL REALITY (VR) AND MARKETING

In today's digital age, it is now possible to imagine a world where everything can be created virtually, at a low cost. These experiences not only provide exposure to alternative realms, but also simultaneously look natural, vibrant, and immeasurable [4,53]. This has all stemmed from the existence of extended reality (XR), and

DOI: 10.1201/9781003196686-4

marketers embracing its use in the social media marketing space, trying to deliver an unusual and unexpected level of experience to customers interacting with their brand.

It was not too long ago that social media only provided the facility to send text messages to its consumers. Now, with the evolution of the world of technology, it is possible for users to upload pictures and videos. In fact, dedicated platforms have been developed, allowing consumers to focus on these specific forms of media (e.g., Instagram and Pinterest for pictures, Instagram and Snapchat for short videos) [6]. These swings have had noticeable effects on the usage of social media, and its consequences, with some researchers suggesting that image-based posts communicate better social presence than plain text (e.g. [7,8]).

It is worth noting is that an overabundance of new technologies in the market proposes that the future of social media will be more sensory-rich [6], and the result of that is nowhere – with the invention and adoption of extended reality (XR) – which has taken an increasingly noticeable place in the real world over the last decade. Products and services devoted to the use of virtual reality (VR) have already been industrialized, and are being used by the day-to-day consumer. Industries such as health, tourism, automotive, entertainment, and sports, to name a few, have used VR in order to create a lasting presence in the mind of today's Internet-savvy consumer.

4.1.2 The Allure of Virtual Reality (VR) to Digital Marketers

Virtual reality (VR) offers a promising market and is aggressively drawing digital marketers to itself. This seems to be more attractive to marketers because it helps them provide an immersive experience to their consumers, for example, the virtual try-on tool provided by Warby Parker, where consumers can try glasses on and see previews for how it would look on their faces [9].

Nowadays, apps and software that stimulate or augment reality are used by over a billion people, and two-thirds of these consumers admit that augmented reality influences their purchase decisions. Gartner states that 84% of customers expressed that they feel like brand experiences are just as significant as the real product or service they are buying, and the same customers were prepared to spend 40% more when brands engaged with them via social media [9].

One can define VR marketing as companies using XR technology to promote themselves and their products [10]. VR facilitates an efficient platform for experience/experiential marketing [11–12]. In experiential marketing, companies emphasize developing and delivering remarkable events for customers, which in themselves become products (the "experience"), with the financial value being measured through the transformational aid delivered by these experiences [13].

Another important area for the application of VR is brand management. Companies are incorporating this futuristic concept in order to build trust and relationships with their customers. This use of this technology relies on the generation of an imitation of a true-to-life environment. Sometimes, VR marketing is simply visual, and sometimes it is multi-sensory, depending on the device. A fully functional VR headset offers a completely immersive experience, while an XR Instagram filter merely superimposes entities or images onto the objects captured by the camera.

The concept of VR helps companies create a direct experience environment for their consumers, and is hence most widely popular in the e-commerce sector. One of the most vital aims of any e-commerce retailer is to facilitate the finest possible shopping experience for their buyers, through computer-mediated communication, primarily the Internet [14,15].

4.2 THE CHANGING ERA OF TECHNOLOGY

There was once a time when technology meant stone tools and implements; that would evolve into newer forms every few millennia. However, since the invention of the first computer, we have seen technology continually evolve at an unprecedented and accelerating rate. What took centuries and decades to change, now breaches the frontiers of innovation in less than a year. In the last decade, a clear emphasis has been laid on softening the interface between humans and technology; whether in the form of artificial intelligence, 3D printing, autonomous vehicles or XR [16]. Novel technology is often said to be created from a combination of existing technologies, or is made possible by them, all with the common goal of enhancing the human experience [17].

4.2.1 THE EVOLUTION OF VIRTUAL REALITY VR

The initiative to interact with the digital world is not a new one. The seed of VR as a concept was sown in the mid-1960s when a researcher [18] presented VR as a window that allows a user's perception of the virtual world to be the same as how they feel, observe, interact and experience the real world. VR is one of the most immersive forms of experience known today. According to Steuer [19] "A *Virtual Reality* is defined as a real or simulated environment in which a perceiver experiences telepresence." Many other authors, Hollebeek et al. [20] have formulated different definitions of VR, and two other widely accepted definitions are presented in Table 4.1.

The aforementioned definitions focus on the most common features of VR, i.e., perception, immersion and interactions. Immersion defines how much the senses of consumers are simulated by a virtual environment, perception describes how similar a customer finds this simulated environment to be to the real world and interaction is determined by the amount of freedom given to the user in the virtual environment. These characteristics largely depend on the efficiency and the accuracy of any technologically simulated environment.

VR is centered around the intent to create a fully digital environment and provide consumers with immersive experiences, which are both exact replicas of a realistic experience and simultaneously different from reality, at any point in time. As early as the 1990s, Milgram and Kishino wrote about a reality-virtuality continuum, with the real environment at one extremity, and a fully virtual environment at the other [25]. There is a nuance to understanding the various elements on this continuum, especially in the context of what we today call augmented reality (AR), mixed reality (MR) and virtual reality (VR), as shown in Figure 4.1. AR can be defined as a "medium in which digital information is overlaid on the physical world that is in both spatial and temporal registration with the physical world and that is interactive in time" [26].

TABLE 4.1
Definitions of Virtual Reality

S.No.	Publication	Definition of VR
1	[21]	"The application of three-dimensional computer technology to generate a virtual environment within which users navigate and interact."
2	[22]	"An immersive computing technology (ICT) that incorporates "a set of technologies that enable people to immersively experience a world beyond reality."
3	[23]	"The computer-generated simulation of a three-dimensional image or environment that can be interacted with in a seemingly real or physical way by a person using special electronic equipment (e.g., a helmet with a screen inside, or gloves fitted with sensors)."
4	[24]	"[When] a user's movements are tracked and his or her surroundings rendered, or digitally composed and displayed to the senses, in accordance with those movements...Substitut[ing] our physical environment and our sensory experiences - what we understand as reality - with digital creations."

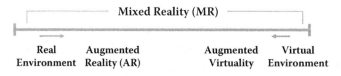

Reality-Virtuality (RV) Continuum

FIGURE 4.1   Milgram and Kishino's Mixed Reality on the Reality-Virtuality Continuum.

Source: Milgram, P., Takemura, H., Milgram and Kishino [25]. Augmented reality: a class of displays on the reality-virtuality continuum. Proc. SPIE 2351, 282–292. doi: 10.1117/12.197321.

MR is everything that lies on the spectrum between VR and the real environment around us. According to K. Dwivedi and Ismagilova [27] there have also been further classifications of MR into AR and augmented virtuality (AV).

VR systems can further be broadly categorized based on the *degree of immersion* presented to the consumers [28], such as:

i. Non-immersive Systems: In this category of VR, users interact with characters or activities present in the virtual world while being physically present in the real world. The motion and actions of the user in the real world are mirrored by those in the virtual world. The most common example of such systems is the world of virtual gaming through consoles such as Sony's PlayStation, Microsoft's Xbox, Nintendo's Wii Sports, etc. Although in such systems, users interact with the virtual world but are not the centers of attention.

ii. Fully Immersive Systems: Unlike a non-immersive environment, a fully immersive environment ensures that the virtual environment provides a convincing realistic experience to the consumer. It attempts to make users feel like they are physically present in the virtual environment, and that all the interactions they are having are real. It is highly expensive to create a VR system that provides complete immersion using the support of various sensory input and output devices such as sense detector helmet, gloves and bodysuits. The movement of the user's head is sensed using a head-mounted display to improve the stereoscopic view of the VR environment. All movements of users in the real world, such as hand movements, the blink of eye, head rotation and bending etc., are taken as inputs through sensors and projected into the virtual world.

iii. Semi-immersive Systems: Semi-immersive systems lie in between non-immersive and fully immersive VR systems. They provide the user with a three-dimensional scene using a monitor and various perspective projections that work in accordance with the movement of the user's head. In this kind of system, the user is the centre of attention. However, their experience of this virtual world is limited to a solely visual experience, with no physical movement involved. Digital movement can be executed using a touch screen on a computer system, or by swiping through mobile systems. Semi-immersive systems are more cost-effective compared to fully immersive systems, and many organizations prefer these to other forms of VR.

4.2.2 EXTENDED REALITY – THE CURRENT TECHNOLOGICAL LANDSCAPE

Much like the technology itself, XR devices have evolved drastically in the last few decades as well, further accelerated by the growth of artificial intelligence (AI) and smartphone technology [29]. Ranging from Facebook-owned Oculus to Sony's Playstation VR, VR consumer devices are changing the game rapidly, amidst stiff competition [30]. AR devices on the other hand can range from the rudimentary HUDs (heads-up displays) widely used across the automobile sector, holographic displays, handheld devices such as smartphones, to more complex smart glasses like Google Glass, Microsoft Hololens or Snap Spectacles [11]. Because of the limitations of today's technology, the term generally used to refer to smart glass technology is "assisted reality" [27]. These devices, though fast-evolving, are currently only useful to effectively display limited amounts of information, usually text, and two-dimensional images.

VR technology involves the use of high-end devices that are vital to the creation and facilitation of immersive experiences in virtual environments. In various related literature work [31], these devices have been broadly categorized into two kinds: input and output devices.

4.2.2.1 Input Devices for VR

Input devices provide users with the capability to interact with VR systems. There could be a variety of input devices such as keyboards, gloves fitted with sensors, and joysticks that map the movements of fingers or postures. These devices are simple to use and easily project discrete commands for a motion to the virtual environments. Bodysuits and gloves equipped with bend sensors can capture a user's motion and their complete hand gestures. These captured movements in the real world are then translated into inputs for the virtual world to imitate the experience.

4.2.2.2   Output Devices for VR

Output devices try to create an immersive experience by allowing the customer to feel, touch, smell, or hear whatever is happening in the virtual world. There is a vast range of output devices, but these are too expensive to be adopted by all scales of businesses and/or consumers. These output devices could be as inexpensive as desktop screens, or as expensive as head-mounted devices, CAVE (Cave Automatic Virtual Environment), or VR glasses. There is a category of output devices known as *haptic devices* that stimulate the senses of the users and allow them to experience the feeling of touch, smell, force, etc.

The applications of XR are neither limited by technological advancements, nor by industry, and this market is expected to grow almost tenfold in the next four years to nearly USD$300 billion [32]. Gone are the days when the only consumer-centric applications of VR were limited to the gaming industry.

What has made XR technology so exciting to both technology producers and consumers alike is its versatility and the diverse potential of applications that exist for each of its forms, i.e., VR, AR and MR. It might not be long before the universe of XR becomes just as indispensable to human interaction as smartphones are today. It is no surprise, therefore, that there has been an increased emphasis on research towards the need to understand how, and to what extent, brands can best utilize XR [27]. Customers today demand more immersive experiences, greater accessibility, and absolute transparency – all at the tip of their fingers (or with a blink of their eyes).

4.2.3   VIRTUAL REALITY AND THE COVID-19 PANDEMIC

The year 2020 saw the world spin into a crisis, with the COVID-19 pandemic, disrupting in its wake, all things technology. As we continue to reel from the effects of the pandemic, and vaccination attempts are underway, the transformative influence of COVID-19 remains undeniable [33,34]. Both early adopters and laggards in the world of XR are making massive moves towards understanding the implications of the onset of restricted mobility, trans-geographic remote collaboration, and communication – often without choice. One could argue that the global pandemic seems to have forced the hands of companies, pushing them towards innovation that is not just limited to the functions of sales and marketing but has led them to completely rethink how they view their core business activities as well, especially in consumer-centric businesses.

XR is proving to be more than a mere aid to doing business, but an extension of the business itself, especially with regards to AR applications. True VR, on the other hand, is centered around "telepresence," which is being employed to enhance experiences by substituting in-person interaction [19]. While consumers demand the same amount of attention as they always have, the means now at a brand's disposal, extend beyond the realm of reality. There were ~5.5 million VR headsets sold in 2020, and this number is expected to grow to over 40 million by 2025 [35]. However, it is important to note that this number is less than 1% of the total number of internet users in the world [36]. AR, on the other hand, while much more accessible, and with much greater reach, might not offer the extent of immersion that consumers are looking for, and might end up looking like no more than a gimmick. Brands must therefore choose wisely, and understand what would the best path forward be while entering into the realm of XR.

4.3 THE EVOLUTION OF MARKETING STRATEGIES

Dr. Philip Kotler defines marketing as

> "The science and art of exploring, creating, and delivering value to satisfy the needs of a target market at a profit [37]. Marketing identifies unfulfilled needs and desires. It defines, measures and quantifies the size of the identified market and the profit potential. It pinpoints which segments the company is capable of serving best and it designs and promotes the appropriate products and services."

Traditional marketing methods involved commercials on the radio and television, large banner ads on billboards, paid content in newspapers and magazines, or coupons and postcards mailed directly to target consumers. Though focus may have slowly been waning away from these methods, they are not yet obsolete, especially when local businesses are trying to reach isolated target audiences without competing against larger businesses.

With the emergence of the Internet, digital marketing has enabled businesses to dip into a practically infinite audience, with minimal financial investments. Moreover, digital marketing has opened the doors to previously unattainable analytics, enabling businesses to make informed decisions about the audience they are selling to. With the rate at which Internet-based technology evolves, digital marketers need constantly, to be ahead of the curve, in order to remain relevant.

4.3.1 THE DIGITAL ERA OF SOCIAL MEDIA MARKETING

The last two decades have witnessed the further evolution of online interactions into social media. Social media, as defined by M. Kaplan and Haenlein [38] is "a group of Internet-based applications that build on the ideological and technological foundations of Web 2.0, and allow the creation and exchange of user-generated content." A social network can be identified by a few key features namely – a set of user profiles, a means for users (and businesses) to interact with each other privately or in groups, and create and curate content. The growth of social media has shifted from a simple platform to connect with friends and family online, to a marketplace of products, services, and information. Not only do users interact with each other, and with brands, but they also form communities around brands and content creators. Therefore, the flow of information is no longer unilateral (from producer to consumer). Users have the power to create content about brands, and even alter how a brand is perceived. Such instances of users creating branding content can either be unpaid and organic or through paid collaboration. Estimates say that the world of influencer marketing might be worth up to USD$15 billion by 2022, and is a clear signal towards the attention that brands are paying to social media marketing today [39].

Sellers have set up brand accounts and stores on platforms like Facebook and Instagram, reaching an ever-growing audience of social media users from their target demographic. Marketers now have a world of boundless possibilities and challenges as customers who use social media can no longer be expected to consume content that they do not specifically demand, making it essential to stimulate demand [40]. This may sound like a herculean task, but it is simplified by the sheer volume of data made available to marketers through social networks. With over 5 billion unique mobile device users worldwide, users have near-constant access to social media and therefore are continually exposed to digital marketers [36]. Digital marketers need to continually upskill in order to keep pace with the advancements in social media.

4.3.2 BRIDGING GAPS IN MARKETING THROUGH VIRTUAL REALITY

In the past, the only major shortcoming of digital marketing has been that some consumers were extremely particular about the physical experience associated with buying. This too has changed drastically with the advent of VR technology, enabling users to engage with virtual surroundings in both a visual and tactile manner. While some marketers were quick to hop on to the XR bandwagon and others came trailing behind, there is enough room for everyone to choose between going the VR or the AR route, and molding their marketing strategies accordingly. Each brand needs to carefully assess what its target buyer persona is, and then personalize the experience to make it appeal to a broad, yet relevant, user base. While defining their marketing mix, marketers need to address all relevant questions related to the use of XR in conjunction with the 4 Ps of marketing [27]. It is safe to say that irrespective of the route a brand takes to get there, most leading brands, whether B2B or B2C, will have to engage with the medium of XR in order to reach, retain and grow their consumer base.

4.4 APPLICATIONS OF EXTENDED REALITY – COMPANIES BEING INNOVATORS

The expanding realm of XR has seen an increased acceptance over the last few years, across industries [41]. Leading automobile manufacturers and sellers like BMW and Vroom are using VR to save costs on prototyping, and AR to enhance the customer experience – bringing the showroom to buyers' hand-held devices [42]. At the Tokyo motor show, Nissan used Oculus VR to allow consumers to design their own Nissan car ([13]). Other European automotive manufacturers like Jaguar are no strangers to the use of VR in-vehicle launches either [43].

Meanwhile, in the healthcare space, VR is being increasingly utilized in medical training and diagnosis as well as in the assessment, understanding, and treatment of mental health disorders [44]. Other industries that are actively employing XR techniques are retail, real estate and architecture, design, L&D and recruitment, education, gaming, and entertainment [45].

4.4.1 Industries and Their Adoption of Extended Reality

Consumers today demand not just features and benefits, but also an emotional connection to the product or service they are engaging with. The field of marketing today is not limited by functionality, but by the experience orientation of the brand [46]. Coca-Cola is one such company, pushing the envelope on experiential marketing, having experimented with multiple VR and AR applications over the years. Coca-Cola collaborated with FIFA for the 2018 Football World Cup, enabling viewers across the globe to engage in a 360-degree tour of the plane that carried the World Cup trophy around the world [47]. The sports and entertainment industry doesn't lag far behind, with football clubs like Manchester City FC launching AR stadium tours for fans, and entertainment organizations like iQIYI, Inc. organizing an immersive reality concert for fans worldwide [48].

In a post–COVID-19 world, with social interaction redefined, it's no surprise that organizations are making an increased push towards adopting XR. But there is a clear edge that the early adopters have already conceptualized XR, especially in sectors like retail and consumer goods. XR can have applications in two broad contexts: (i) remote interaction and (ii) in-person interaction. Remote interaction is when users engage with XR from their personal devices without directly interacting with the brand. On the other hand, in-person interaction implies a direct physical interaction with the brand, with XR being used to improve customer experience. Figure 4.2 examines our understanding of which form of interaction a few key industries could be expected to have with XR technology, along with the impact, whether positive (+ve) or negative (–ve) that the COVID-19 pandemic has had on them [49].

4.4.2 The Use of Extended Reality in Marketing

The most intriguing applications of XR are undeniably in the social space. Enhancing the engagement duration is one of the central goals for any digital marketer. Fifty-six percent of a brand's sales lift from digital advertising can be attributed to the quality of the creative according to a Nielsen Catalina survey [50]. It is thus clear that with an average dwell time of 75 seconds, marketing campaigns that utilize AR are here to stay [51] The cost-friendly and mainstream nature of VR [52] has led to social media playing a vital role in its mounting usage. Marketers are now integrating AR into their social media platforms like Instagram and Snapchat.

AR is one of the most prominent technological innovations that has already dug its claws into social media. Possibly the most familiar examples of this would be

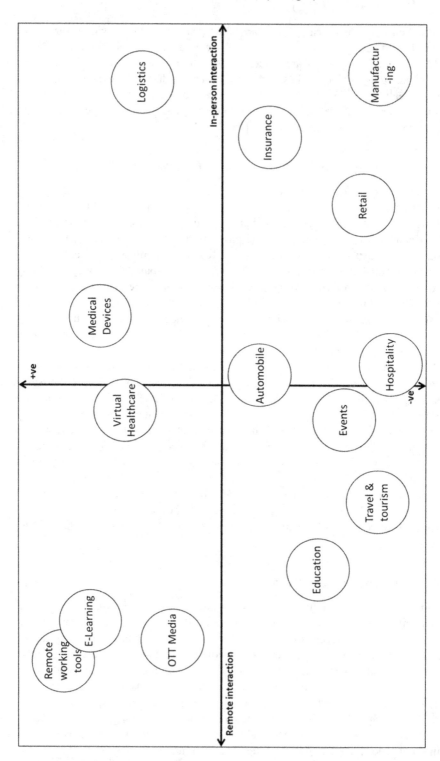

FIGURE 4.2 A Graphical Representation of the Expected Adoption of Extended Reality by Selected Industries Post–COVID-19.

Snapchat's filters, which use a mobile device's camera to overlay real-time visual and/ or video, superimposed on people's faces (including features such as animal ears makeup, jewelry, etc.) [6]. The company has even launched filters to explicitly be used on users' cats [53]. Other marketers dealing in social media speedily joined the AR movement, to name a few, Instagram's current adoption of AR filters [54], and Memoji messaging started by the well-known consumer electronics brand, Apple [55].

4.4.2.1   Extended Reality in the Consumer Electronics Industry

Some brands use more advanced features to superimpose brand personas or products onto the real world. Leading smartphone manufacturer, Oneplus, successfully launched the Oneplus Nord at an AR event, allowing users to interact with, and manipulate the device virtually [56]. Oneplus has also experimented with XR through advertisements on Instagram, allowing users to view a virtual rendering of a Oneplus device in 3D.

4.4.2.2   Extended Reality in the Beauty and Fashion Industries

Beauty brands aren't far behind either, with the likes of Mac cosmetics creating AR filters for makeovers, and Sephora setting up an AR shopping experience with real-time facial recognition while also employing Facebook's AR ads services [57–58]. Footwear brands such as Gucci and Nike allow users to try on shoes virtually, before purchase, to alleviate fitting concerns [59,60]. On the other hand, in the eyewear space, brands like Michael Kors and Ray-ban have also ventured into this space [61]. Popular fashion brands like Asos, are allowing consumers to virtually try on clothing on their websites, applications, and through direct ads on users' Snapchat, Facebook news feeds or Instagram feeds and stories – for the ideal fit [62].

4.4.2.3   Extended Reality in the Retail and E-commerce Industries

Another example of the use of VR by companies for relationship building is that of Boursin Cheese's "Sensorium," which permits customers in shopping centers to see inside the contents of a fridge. E-commerce has grown to a global scale because of superior Internet access, search engines, and diverse social media systems, such as aggregators, for instance, Kayak.com [63], online consumer reviews [64], and social networks [e.g., Facebook, Instagram [65]]. One of the major limitations of e-commerce has always been that consumers cannot experience products, services, and stores in the same way as they can have multisensory interactions with physical stores [66,67]. With the advent of XR technology, it is possible for e-retailers to create a virtually real shopping experience for their consumer, often referred to as virtual commerce or V-commerce [68]. Through this technology, online shopping experiences have grown from traditional dragging and dropping items into a cart on two-dimensional websites, into a real-time, immersive experience, where users can navigate through virtual stores, and interact with virtual forms of physical products and salespersons, just as they would in physical stores [69].

4.4.2.4   Extended Reality in the Real Estate Industry

Industries severely affected by the COVID-19 pandemic also seem to have found some respite in XR-based applications. The real estate industry was severely affected by the pandemic. The real estate industry is not only utilizing MR applications to advertise projects, but also to facilitate virtual home visits for new buyers [70]. Furnishing and decor brands like IKEA and Home Depot too have launched their own AR platforms for potential buyers to visualize true-to-scale models of furniture before purchase.

4.4.2.5   Extended Reality in the Hospitality Industry

In the hospitality industry, hotels are trying to entice guests through AR experiences within properties. Holiday Inn allows guests to use their smartphones to see virtual renderings of celebrities right in their hotel rooms. Best Western hotels targeted families, allowing kids to view their favorite Disney characters real-time. Hotels have also used AR applications to enable virtual redecoration of rooms, and we might soon see instances where hotels offer advertisements where a smartphone user can point their phone at a digital or physical ad and take a walk through the rooms and facilities of premium hotels and resorts [71,72].

4.4.2.6   Extended Reality in the Travel and Tourism Industry

The travel and tourism industry on the other hand has been experimenting with XR applications for over 5 years, with Thomas Cook's VR "Try Before You Fly" experience that allows customers to experience a series of virtual holidays, or Google Expeditions, which partners with the likes of National Geographic and the American Museum of Natural History to deliver both AR and VR experiences to make travel more accessible [73,74].

4.4.3   Immersive Virtual Reality and Social Networks

Virtual reality social networks are indeed the most novel and immersive experiences available today. Platforms like Oculus Rooms, Facebook Horizon, or AltspaceVR are at the frontier of technological innovation, offering collaborative virtual spaces for users to interact, work and play together in a purely virtual environment [75–76]. Time zones and geography are no longer a concern for working professionals looking to collaborate from across the world, or for the casual user looking to connect with a distant relative. As this technology continues to evolve, there are ethical questions around privacy and autonomy that these companies need to address [77].

4.5   THE FUTURE OF VIRTUAL REALITY AND SOCIAL MEDIA MARKETING

The future of social media marketing and that of VR seems to be intertwined. Technology is no longer restricted to Millennial and Generation Z users but finds more relevant than ever at the workplace. Tasks that would earlier require people to travel halfway across the world to collaborate on, are completed just as effectively, and even more efficiently, using VR. Social VR applications are at a very nascent stage of development and show immense promise from the perspective of enabling presence [1].

The term *metaverse* could be used to refer to an interconnected web of XR environments, stitched together throughout our real world [78]. From being able to shop for groceries virtual vending machines to be directly delivered to your doorstep, to interacting with three-dimensional monthly expense reports at work, the metaverse may not be as distant a future as it was once thought to be. The next revolution after social media will indeed be that of the metaverse, where marketers will have countless unexplored avenues to advertise to both a consumer and their digital twin [1].

Taking a step further, and crossing the frontiers of the metaverse, one might find a world of infinite possibilities. Imagine dating apps of the future, where you could interact with near-real avatars of your matches, from anywhere in the world. You could then travel to your favorite virtual fashion retail store and rent out your ideal dress with virtual currency. Once you're both ready to go, you could head out on a virtual date to a romantic destination of your choosing, and eat at a near-real virtual rendering of your favorite restaurant in the world. Couples could travel the virtual world together, exploring each other's dream destinations, before booking tickets to meet at their chosen destination in the real world. It might not be a stretch of the imagination to say that we might also have a purely virtual workforce, with architects working alongside artificial intelligence to build monuments in this imaginary world. In fact, if technology evolves to the extent where the human need for emotional companionship can be met in VR, it could completely transform the way social relationships like marriages and families exist; with two people from diverse ethnoreligious backgrounds uniting in a legally binding virtual matrimony.

Though we might still be a long way from a virtual utopia as immersive as Ernest Cline's fictional world – The Oasis – from Ready Player One, we need to be prepared for a VR not much different from it. The primary limitation to a virtual world as advanced as the Oasis lies not in the visual, but merely in the tactile stimulus required to operate in such an environment [79].

4.6 CONCLUSION

With social media headed towards an ecosystem based on the interfaces between the brains of users, and their computers, a virtual world may soon become a reality for many. If marketers have learned anything from the COVID-19 pandemic, it is that presence cannot always be limited to the physical self, and are thus looking to find the most optimal ways to ensure their brands do not lose appeal. The above study touches the multidisciplinary areas because fragmented filed are involved such as marketing, social media marketing, technology evolution, and the increasing significance of virtual reality. Virtual reality and augmented reality are identified as analytical digital technologies that will lead to novel marketing prospects to create a distinct image in the minds of consumers. The majority of the work is being done in the area of augmented reality as it is affordable and in access to many companies, but virtual reality is creating new hopes for companies particularly in the pandemic situation. Virtual stores are going to give tough competition to the physical stores provided the experiences seem similar as well as fascinating to the consumer.

A detailed study of the role of XR, VR, AR, and MR shows there is definite potential for advancement, both in the technology and its application. Trends suggest a greater acceptance and shift towards such applications, with business users and marketers across industries finding increasingly relevant ways to use this technology to their advantage. The adoption of XR technology also hints towards the potential of this virtual world expanding beyond being an aid to product marketing, but a product line in itself. Targeted marketing and advertising can be expected to evolve beyond our imaginations once social life extends into the metaverse [1]. The COVID-19 pandemic has set the wheels in motion for additional research on the subject, especially in those industries most drastically hit, and a lot remains to be seen with how the marketing landscape of the future truly evolves. For brands that are willing to take the leap, there could not be a better time. As we stand at the cusp of another digital revolution, marketers can finally deliver to customers their wildest imaginations [4].

REFERENCES

[1] Hackl, C. (2020, August 30). *Social VR, Facebook Horizon and the Future of Social Media Marketing*. Retrieved April 10, 2021, from Forbes: https://www.forbes.com/sites/cathyhackl/2020/08/30/social-vr-facebook-horizon--the-future-of-social-media-marketing/?sh=8216b5a5b19c

[2] Hackl, C. (2020, July 5). *The Metaverse Is Coming and It's a Very Big Deal*. Retrieved April 10, 2021, from Forbes: https://www.forbes.com/sites/cathyhackl/2020/07/05/the-metaverse-is-coming--its-a-very-big-deal/?sh=2ec8952e440f

[3] Statista. (2021, March 17). *Consumer and Enterprise Virtual Reality (VR) Market Revenue Worldwide from 2019 to 2024*. Retrieved April 1, 2021, from statista: https://www.statista.com/statistics/1221522/virtual-reality-market-size-worldwide/

[4] Stuart, H. (2019). *Virtual Reality Marketing: Using VR to Grow a Brand and Create Impact*. London, United Kingdom: KogenPage.

[5] Appel, G., Grewal, L., Hadi, R., & Stephen, A. T. (2019, October 12). The future of social media in marketing. *Journal of the Academy of Marketing Science* (48), 79–95.

[6] Pittman, M., & Reich, B. (2016). Social media and loneliness: Why an Instagram picture may be worth more than a thousand twitter words. *Computers in Human Behavior*, 62, 155–167.

[7] Piumsomboon, T., Day, A., Ens, B., & Billinghurst, M. (2017, November). *Exploring Enhancements for Remote Mixed Reality Collaboration*. Retrieved April 13, 2021, from Researchgate: https://www.researchgate.net/publication/321405854_Exploring_enhancements_for_remote_mixed_reality_collaboration

[8] Bdex. (n.a.). *Immersive, Augmented, and Virtual Realities Are Changing the Social Media Landscape*. Retrieved March 29, 2021, from Bdex.com: https://www.bdex.com/ar-vr-social-media-marketing/

[9] Adespresso. (2020, September 17). *How Virtual Reality Marketing Is Changing the Face of Consumer-Brand Interaction*. Retrieved April 2, 2021, from adespresso.com: https://adespresso.com/blog/virtual-reality-marketing/

[10] Pine, B. J., & Gilmore, J. H. (1998). Welcome to the experience economy. *Harvard Business Review*, 76(4), 97–105.

[11] Software Testing Help. (2021, March 27). *10 BEST Augmented Reality Glasses (Smart Glasses) In 2021*. Retrieved April 12, 2021, from Software Testing Help: https://www.softwaretestinghelp.com/best-augmented-reality-glasses/

[12] Barnes, S. J. (n.a.). *Understanding Virtual Reality in Marketing: Nature, Implications and Potential*. London: King's College London, Department of Management.

[13] Dennis, C., Fenech, T., Pantano, E., Gerlach, S., & Merrilees, B. (2004). *E- Retailing*. Abingdon: Routledge.

[14] Hoffman, D. L., & Novak, T. P. (1996). Marketing in hypermedia computer-mediated environments: Conceptual foundations. *Journal of Marketing*. 60, 50–68. https://doi.org/10.1177/002224299606000304

[15] Kenwright, B. (2020, January). *The Future of Extended Reality (XR)*. Retrieved April 12, 2021, from Virtual Reality: Ethical Challenges and Dangers: https://www.xbdev.net/misc_demos/demos/future-of-xr/paper.pdf

[16] Arthur, W. B. (2009, August). *The Nature of Technology: What It Is and How It Evolves*. Retrieved April 12, 2021, from Simon and Schuster: https://books.google.co.in/books?hl=en&lr=&id=3qHs-XYXN0EC&oi=fnd&pg=PA1&dq=why+technology+evolves&ots=5ZNhoN7VBa&sig=dpiso2U9WsAzcfhL2xH_JMIZUkI&redir_esc=y#v=onepage&q=why%20technology%20evolves&f=false

[17] Sutherland, I. E., & Sutherland, I. E. (1965). The ultimate display. *Proceedings of the IFIP Congress*, 2, 506–508. Retrieved April 02, 2021, from http://citeseer.ist.psu.edu/viewdoc/summary?doi=10.1.1.136.3720

[18] Steuer, J. (2006, February 7). *Defining Virtual Reality: Dimensions Determining Telepresence*. Retrieved April 12, 2021, from International Communication Association - Journal of Communication: https://academic.oup.com/joc/article-abstract/42/4/73/4210117?redirectedFrom=fulltext

[19] Hollebeek, L. D., Clark, M. K., Andreassen, T. W., Sigurdsson, V., & Smith, D. (2020). Virtual reality through the customer journey: Framework and propositions. *Journal of Retailing and Consumer Services*, 55, 102056. https://doi.org/10.1016/j.jretconser.2020.102056

[20] Cowan, K., & Ketron, S. (2019). A dual model of product involvement for effective virtual reality: The roles of imagination, co-creation, telepresence, and interactivity. *Journal of Business Research*, 100, 483–492. https://doi.org/10.1016/j.jbusres.2018.10.063

[21] Berg, Leif P., & Vance Judy M. (2017). Industry use of virtual reality in product design and manufacturing: A survey. *Mechanical Engineering Publications*, 205. https://lib.dr.iastate.edu/me_pubs/205

[22] KPMG (2016). *How Augmented and Virtual Reality Are Changing the Insurance Landscape*. Retrieved Oct 17, 2018, from https://assets.kpmg.com/content/dam/kpmg/xx/pdf/ 2016/10/how-augmented-and-virtual-reality-changing-insurance-landscape.pdf

[23] Fox, J., Arena, D., & Bailenson, J. (2009). Virtual reality: A survival guide for the social scientist. *Journal of Media Psychology*, 21(3), 95–113.

[24] Milgram, P., & Kishino, F. (1994, December). *A Taxonomy of Mixed Reality Visual Displays*. Retrieved April 12, 2021, from ResearchGate: https://www.researchgate.net/publication/231514051_A_Taxonomy_of_Mixed_Reality_Visual_Displays

[25] Craig, A. B. (2013). "What is augmented reality?" in A. B. Craig (Ed.), *Understanding Augmented Reality: Concepts and Applications* (p. 20). Waltham: Elsevier. Retrieved from https://books.google.co.in/books?hl=en&lr=&id=7_O5LaIC0SwC&oi=fnd&pg=PP1&ots=LHFDt_uPqa&sig=9Bc9ngBrBnGsvNcObtqOhJzMHLs&redir_esc=y#v=onepage&q&f=false

[26] K. Dwivedi, Y., & Ismagilova, E. et al. (2020). Setting the future of digital and social media marketing research: Perspectives and research propositions. *International Journal of Information Management*.

[27] Ventura, S., Brivio, E., Riva, G., & Baños, R. M. (2019). Immersive versus non-immersive experience: Exploring the feasibility of memory assessment through 360° technology. *Frontiers in Psychology*, 10, 2509. https://doi.org/10.3389/fpsyg.2 019.02509

[28] Luck, M., & Aylett, R. (2010, November 26). Applying artificial intelligence to virtual reality: Intelligent virtual environments. *Taylor and Francis Online*, 3–32. Retrieved April 13, 2021, from Taylor and Francis Online-Applied Artificial Intelligence: https://www.tandfonline.com/doi/abs/10.1080/088395100117142

[29] Stein, S. (2021, April 4). *The Best VR Headset for 2021*. Retrieved April 12, 2021, from CNET News: https://www.cnet.com/news/the-best-vr-headset-for-2021/

[30] Cipresso, P., Giglioli, I. A. C., Raya, M. A., & Riva, G. (2018). The past, present, and future of virtual and augmented reality research: A network and cluster analysis of the literature. *Frontiers in Psychology*, 9(Nov), 2086. https://doi.org/10.3389/ fpsyg.2018.02086

[31] Boston Consulting Group. (2021). *Making Augmented Reality and Virtual Reality Your Business Reality*. Retrieved April 12, 2021, from Boston Consulting Group-Digital, Technology, and Data: https://www.bcg.com/capabilities/digital-technology-data/emerging-technologies/augmented-virtual-reality

[32] Matthews, B., See, Z. S., & Day, J. (2020, October 20). *Crisis and Extended Realities: Remote Presence in the Time of COVID-19*. Retrieved April 12, 2021, from Sage Journals: https://journals.sagepub.com/doi/full/10.1177/1329878X20967165

[33] Meißner, M., Pfeiffer, J., Pfeiffer, T., & Oppewal, H. (2019). Combining virtual reality and mobile eye tracking to provide a naturalistic experimental environment for shopper research. *Journal of Business Research*, 100, 445–458. https://doi.org/1 0.1016/j.jbusres.2017.09.028

[34] Vailshery, L. S. (2021, March 16). *Augmented Reality (AR) and Virtual Reality (VR) Headset Shipments Worldwide from 2020 to 2025*. Retrieved April 12, 2021, from Statista: https://www.statista.com/statistics/653390/worldwide-virtual-and-augmented-reality-headset-shipments/#:~:text=Estimates%20suggest%20that%20in%202020, units%20per%20year%20by%202023

[35] Data Portal. (2021, March). *Digital Around the World*. Retrieved April 12, 2021, from Data Portal: https://datareportal.com/global-digital-overview#:~:text=The %20number%20of%20internet%20users,875%2C000%20new%20users%20each %20day

[36] Kotler Marketing Group. (2019). *Dr. Philip Kotler Answers Your Questions on Marketing*. Retrieved April 10, 2021, from Kotler Marketing Group: https://www. kotlermarketing.com/phil_questions.shtml

[37] M. Kaplan, A., & Haenlein, M. (2010). Users of the world, unite! The challenges and opportunities of Social Media. *Elsevier- Business Horizons*, 53(1), 59–68.

[38] Insider Intelligence. (2021, January 6). *Influencer Marketing: Social media influencer market stats and research for 2021*. Retrieved April 10, 2021, from Insider Intelligence, 59, 15–16. https://www.businessinsider.com/influencer-marketing-report? IR=T

[39] Guzzo, T., Andrea, A. D., & Ferri, F. (2012). Evolution of marketing strategies: From internet marketing to M-marketing. *OTM Confederated International Conferences "On the Move to Meaningful Internet Systems"* (pp. 627–636). Berlin: Springer.

[40] Terra, D. (2017). *Virtual Reality in the Automotive Industry*. Retrieved April 12, 2021, from Developers: https://www.toptal.com/virtual-reality/virtual-reality-in-the-automotive-industry

[41] Businesswire. (2016, August 2). *Vroom Launches First-Ever Dynamic Virtual Reality Car Showroom, Continuing to Pave the Way for Future of Automotive E-Commerce.* Retrieved April 13, 2021, from Businesswire: https://www.businesswire.com/news/home/20160802005105/en/Vroom-Launches-First-Ever-Dynamic-Virtual-Reality-Car-Showroom-Continuing-to-Pave-the-Way-for-Future-of-Automotive-E-Commerce

[42] Jaguar. (2017). *Jaguar I Pace – Concept Launch.* Retrieved April 12, 2021, from Rewind: https://rewind.co/portfolio/jaguar-i-pace-concept-launch/

[43] Joshi, N. (2021, April 4). *Exploring the Applications of Extended Reality in Heathcare, Real Estate and Retail.* Retrieved April 12, 2021, from BBN Times: https://www.bbntimes.com/technology/exploring-the-applications-of-extended-reality-in-healthcare-real-estate-and-retail

[44] Thompson, S. (2020, December 11). *VR Applications: 21 Industries Already Using Virtual Reality.* Retrieved April 12, 2021, from Virtual Speech: https://virtualspeech.com/blog/vr-applications

[45] Schmitt, B. (1999). Experiential marketing. *Journal of Marketing Management,* 15, 53–67.

[46] Erang. (2018, June 27). *Immersive Campaign of the Year With Coca-Cola & FIFA.* Retrieved April 12, 2021, from ByondXR: https://www.byondxr.com/blog/brands/immersive-promotion-with-coca-cola-fifa-trophy-tours/

[47] Cision PR Newswire. (2021, March 28). *iQIYI Ushers in Next-Generation of Entertainment with Chinese Girl Group THE9's Debut Extended Reality (XR) Concert.* Retrieved April 12, 2021, from Cision PR Newswire: https://www.prnewswire.com/news-releases/iqiyi-ushers-in-next-generation-of-entertainment-with-chinese-girl-group-the9s-debut-extended-reality-xr-concert-301257152.html

[48] Somanas, A. (2020, April 3). *10 Business Sectors Boosted by Coronavirus Concerns.* Retrieved April 12, 2021, from FM Magazine Website: https://www.fm-magazine.com/news/2020/apr/business-sectors-boosted-by-coronavirus-concerns.html

[49] Perry, E. (2020, January 14). *How Custom Instagram AR Filters Can Boost Your Brand's Personality.* Retrieved April 10, 2021, from Social Media Week: https://socialmediaweek.org/blog/2020/01/how-custom-instagram-ar-filters-can-boost-your-brands-personality/

[50] India, C. (2016, June 16). *What Does the Future Hold for Augmented Reality in Digital Marketing?* Retrieved April 13, 2021, from Rubix: https://rubixmarketing.uk/2018/04/06/augmented-reality-digital-marketing/

[51] Colville W. (2018). *Facebook VR Leader Talk About the Future of Virtual Marketing.* Retrieved from https://tinyurl.com/y8kdd4cr.

[52] Ritschel, C. (2018). *Snapchat Introduces New Filters for Cats.* Retrieved March 31, 2021, from https://tinyurl.com/y8shdhpl

[53] Rao, L. (2017). *Instagram Copies Snapchat Once Again with Face Filters.* Retrieved April 2, 2021, from https://tinyurl.com/ybcuxxdv.

[54] Tillman, M. (2018). *What Are Memoji? How to Create An Animoji That Looks Like You.* Retrieved from https://tinyurl.com/yakqjqdf

[55] Carl. (2020, July 18). *Experience the World's First AR Smartphone Launch.* Retrieved April 10, 2021, from One Plus: https://www.oneplus.in/nord/AR

[56] Facebook. (2018, July 10). *Introducing New Ways to Inspire Holiday Shoppers with Video.* Retrieved April 10, 2021, from Facebook for Business: https://www.facebook.com/business/news/introducing-new-ways-to-inspire-holiday-shoppers-with-video

[57] Naffa, M. (n.d.). *Mac Cosmetics Launches AR Makeovers*. Retrieved April 10, 2021, from Grazia: https://graziamagazine.com/me/articles/mac-cosmetics-launches-augmented-reality-makeovers/

[58] Hobbs, R. (2020, July 15). *Gucci Partners with Snapchat for AR Try-On*. Retrieved April 6, 2021, from Stylus: https://www.stylus.com/gucci-partners-with-snapchat-for-ar-tryon#:~:text=Gucci%20has%20adopted%20Snapchat%27s%20AR,the%20primarily%20Gen%20Z%20platform

[59] Mullan, L. (2020, May 16). *Nike Unveils New Augmented Reality Technology To Improve Shoe Sizing*. Retrieved April 10, 2021, from Technology Magazine: https://www.technologymagazine.com/data-and-data-analytics/nike-unveils-new-augmented-reality-technology-improve-shoe-sizing

[60] Facebook Augmented Reality. (2021). *Michael Kors- Personalizing the shopping experience with Facebook Augmented Reality ads*. Retrieved April 12, 2021, from Facebook for Business: https://www.facebook.com/business/success/4-michael-kors

[61] Walk-Morris, T. (2020, May 12). *Asos Debuts AR Tool For Online Shoppers*. Retrieved April 10, 2021, from Retail Dive: https://www.retaildive.com/news/asos-debuts-ar-tool-for-online-shoppers/577679/#:~:text=Asos%20last%20week%20introduced%20augmented,release%20emailed%20to%20Retail%20Dive

[62] Dellarocas, C., Katona, Z., & Rand, W. (2013). Media, aggregators, and the link economy: Strategic hyperlink formation in content networks. *Management Science Letters*, 59, 2360–2379. https://doi.org/10.1287/mnsc.2013.1710

[63] Zhu, F., & Zhang, X. M. (2010). Impact of online consumer reviews on sales: The moderating role of product and consumer characteristics. *Journal of Marketing*, 74(2), 133–148. https://doi.org/10.1509/jmkg.74.2.133

[64] You, Y., Vadakkepatt, G. G., & Joshi, A. M. (2015). A meta-analysis of electronic word-of-mouth elasticity. *Journal of Marketing*, 79, 19–39. https://doi.org/10.1509/jm.14.0169

[65] Bonetti, F., Warnaby, G., & Quinn, L. (2018). "Augmented reality and virtual reality in physical and online retailing: A review, synthesis and research agenda," in T. Jung and M. Tom Dieck (Eds.), *Augmented Reality and Virtual Reality* (pp. 119–132). Cham: Springer. https://doi.org/10.1007/978-3-319-64027-3_9

[66] Lee, K. S., & Tan, S. J. (2003). E-retailing versus physical retailing: A theoretical model and empirical test of consumer choice. *Journal of Business Research*, 56, 877–885.

[67] Nguyen, B., Pantano, E., Dennis, C., & Gerlach, S. (2016). *Internet Retailing and Future Perspectives*. Abingdon: Routledge.

[68] Alcañiz, M., Bigné, E., & Guixeres, A. J. (2019, July 05). Virtual reality in marketing: A framework, review, and research agenda. *Frontieres in Psychology*, 10(1530). https://doi.org/10.3389/fpsyg.2019.01530

[69] Liu, S. (2018, May 15). *Top Examples of Augmented Reality in Real Estate*. Retrieved April 10, 2021, from Propertyme: https://www.propertyme.com.au/blog/industry-news/augmented-reality-real-estate-examples

[70] Kohles, C. (2020, March 10). *Augmented Reality Experiences For Hotels: Innovative AR Ideas for the Lodging Industry*. Retrieved April 10, 2021, from Wikitude: https://www.wikitude.com/blog-augmented-reality-experiences-for-hotels-innovative-ar-ideas-for-the-lodging-industry/

[71] Revfine- Optimising Revenue. (n.d.). *How Augmented Reality Is Transforming the Hospitality Industry*. Retrieved April 10, 2021, from Revfine- Optimising Revenue: https://www.revfine.com/augmented-reality-hospitality-industry/

[72] Google- For Education. (n.d.). *Bring Your Lessons to Life With Expeditions*. Retrieved April 10, 2021, from Google- For Education: https://edu.google.co.in/products/vr-ar/expeditions/

[73] Thomas Cook. (2015). *Thomas Cook Virtual Reality Holiday 'Try Before You Fly'*. Retrieved April 10, 2021, from Visualise: https://visualise.com/case-study/thomas-cook-virtual-holiday

[74] AltspaceVR. (2021). *AltspaceVR*. Retrieved April 10, 2021, from AltspaceVR: https://altvr.com/

[75] Oculus. (2016, December). *Oculus Rooms*. Retrieved April 10, 2021, from Oculus: https://www.oculus.com/experiences/go/1101959559889232/?locale=en_US

[76] O'Brolcháin, F., & Jacquemard, T. (2015, February 1). *The Convergence of Virtual Reality and Social Networks: Threats to Privacy and Autonomy*. Retrieved April 14, 2021, from Springer Link.

[77] Jaynes, C., & Seales, W. (2003, May). The Metaverse: A networked collection of inexpensive, self-configuring, immersive environments. *ACM Digital Library*, pp. 115–124.

[78] Newman, H. (2018, March 31). *Ready Player One' Versus Reality: How Close Are We?* Retrieved April 13, 2021, from Forbes: https://www.forbes.com/sites/hnewman/2018/03/31/ready-player-one-versus-reality-how-close-are-we/?sh=4d38fcd52a01

[79] Jaekel, B. (2017). *Sephora Boosts Augmented Reality Shopping With Real-time Facial Recognition*. Retrieved April 10, 2021, from Retail Dive: https://www.retaildive.com/ex/mobilecommercedaily/sephora-tries-on-augmented-reality-update-for-real-time-facial-recognition#:~:text=The%20augmented%20reality%20feature%20currently,makeup%2C%20with%20more%20effective%20technology

[80] Oculus from Facebook. (n.d.). *Horizon*. Retrieved April 10, 2021, from Oculus from Facebook: https://docs.google.com/document/d/1m8LYiUl67iWfc9lmYHeUyEZpK_TxQ_hcRRgESroh3_0/edit#

5 An Efficient Deep Learning Framework for Multimedia Big Data Analytics

G. S. Pradeep Ghantasala[1], L. R. Sudha[2], T. Veni Priya[2], P. Deepan[2] and R. Raja Vignesh[3]

[1]Institute of Engineering and Technology, Chitkara University, Punjab, India
[2]Annamalai University, Annamalainagar, Chidambaram, Tamilnadu, India
[3]K.S.K. College of Engineering and Technology, Kumbakonam, Tamilnadu, India

5.1 INTRODUCTION

Multimedia data means data from two or more medium. Typically, it represents data from different types of medium and normally utilized types of data are alphanumeric, numbers, text, pictures, sound and video. In like manner use, individuals allude a data when multimedia just when data is time-dependent, for example, sound and video are included. Big data is a word that pronounces a vast quantity of data, which can be structured or unstructured, growing drastically with time. The rapid growth in the computing and mobile technologies, the Internet, social media, storage mediums and the user-generated e-content have a great propensity towards the worldwide generation of tremendous amount of multimedia big data in a variety of formats (like as text, audio, video and animation) and this contributes to the ever-booming universal multimedia data sphere [1]. Multimedia analytics is a new and enormously growing research area that uses ensemble techniques such as visual analytics, data analytics and data management in order to come up with a system that analyse a large volume of multimedia data effectively.

Extracting useful information from large-scale multimedia is a crucial step in multimedia data analytics. The extraction of salient information from multimedia

DOI: 10.1201/9781003196686-5

big data can be done using data-driven and model-centric techniques [2]. In model-centric technique feature extraction is done using human-engineered features and by leaning on human heuristics. Gabor filters, wavelets and scale-invariant feature transforms (SIFT) are some of the most commonly used versions. A foremost shortcoming of the model-centric strategy is it will not consider unusual circumstances like changes due to environmental factor, lighting conditions, etc. during its design. The data-driven technique contradictory to model-centric technique learns representation from data rather than using human-engineered features. The greater the diversity and quantity of data, the better the learning of a data-driven technique.

While thinking about multimedia analytics, one of the most common deep learning techniques is effective techniques that can be used to fetch salient information from multimedia data effectively.

The following are the key obstacles for major multimedia data analytics [3]:

- Due to the variety of modality, fusion of heterogeneous data is possible.
- Extraction of high-level and low-level features for model representation.
- For multimedia big data, deep learning and artificial intelligence techniques are used.
- Scalability, data storage, management and processing of multimedia big data.
- Handing the large amount of data (social media data like Twitter and Facebook).

In order to handle the above-mentioned issues, we have to use artificial intelligence and deep learning techniques. The convergence of deep learning in multimedia analytics has boosted the performance of several multimedia tasks, such as classification, detection and regression, and has also fundamentally changed the landscape of several relatively new areas, such as semantic segmentation, captioning and content generation.

5.2 DEEP LEARNING

A type of machine learning technique known as deep learning is used to build a model that learns from data like images, audio, video or text and performs tasks like classification, identification, etc. The deep learning technique is considered part of machine learning, a subcategory of artificial intelligence [4], and machine learning is a detachment of artificial intelligence. Figure 5.1 depicts the relationship between artificial intelligence, deep learning and machine learning.

In general, deep learning algorithms are implemented using neural network architecture. The term *deep* in deep learning indicates the amount of layers in the network and the network will become deeper with a rise in the number of layers. In general, deep networks can have hundreds of layers, whereas the conventional neural networks contain only two or three layers. The accuracy provided by deep learning techniques has made it a classification and identification methodology that

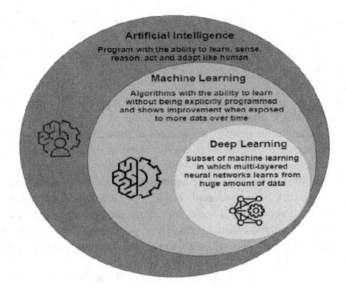

FIGURE 5.1   Artificial Intelligence (AI), Machine Learning (ML) and Deep Learning (DL): A Rapport.

is cutting-edge. The three pillars that make this degree of accuracy possible are easy access to massive sets of labelled data, increase in computing power and the availability of pre-trained models built by experts. The different model of deep learning is depicted in Figure 5.2 and it consists of scene classification and object detection related to multimedia analytics.

5.2.1 BASIC ARCHITECTURE

The neural network architecture entails many layers and the input layers form the first layer of the neural network. The input layer collects the input i, from whichever

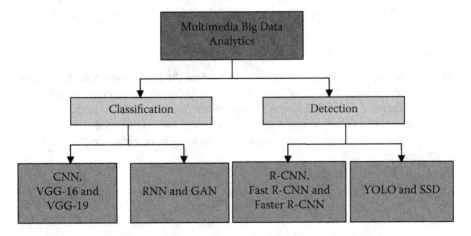

FIGURE 5.2   Deep Learning Models for Scene Classification and Object Detection.

the neural network learns. The input value i is an entire vector and each value in the vector (word or pixel), will be considered as a particular neuron in an neural network and it forms the basic building block of a deep learning neural network. It is considered the perceptron of an artificial neural network [5]. The basic architecture of deep neural networks is shown in Figure 5.3.

The last layer of such a neural network has the output layer, which produces an output vector o, which represents the result produced by the neural network. The entries in the output vector signify the standards of those neurons that have the output layer. In circumstance of organization, every neuron in the output layer represents a diverse class and the significance of a neuron in the output layer offers the possibility of the object specified by the features i is appropriate to one of the probable classes [6].

The neural network has to accomplish a set of mathematical operations in order to obtain the output vector o and these procedures are accomplished by the hidden layers among the input and output layers. The layer connection in a neural network is depicted in Figure 5.4. The network input layer comprises three input neurons, while the output layer consumes four neurons.

Every connection between a pair of neurons is characterized by weight w. The weight w is the numerical value and will be different for each connection. Each of the weights w will have indices. The first index 2 represents the neuron in the layer of which the association starts and the second index represents the neuron in the layer to whichever the association indicates. The weights among the two layers of the neural network could be characterised by a weight matrix, as shown in Equation 5.1.

$$W = \begin{bmatrix} w11 & w12 & w13 & w14 \\ w21 & w22 & w23 & w24 \\ w31 & w32 & w33 & w34 \end{bmatrix} \quad (5.1)$$

The amount of entries in the matrix will be equivalent to the number of connections between two layers. The quantity of rows in the matrix agrees to the quantity of neurons in a layer since when the connection originates, the quantity of columns resembles the amount of neurons in the layer to which the connection leads.

5.2.2 LEARNING PROCESS

The first step in the learning process is to determine the prediction vector h. For any given input vector i, the neural network will calculate the prediction vector h, as shown in Figure 5.5 and this step is known as forward propagation. In this step, the dot product among input vector I and weight matrix w is computed.

The dot product among input vector i and the weight matrix w yields a vector z, as shown below:

$$\vec{i}^T . w = (i_1, i_2). \begin{pmatrix} w_{11} & w_{12} & w_{13} & w_{14} \\ w_{21} & w_{22} & w_{23} & w_{24} \\ w_{31} & w_{32} & w_{33} & w_{34} \end{pmatrix} \quad (5.2)$$

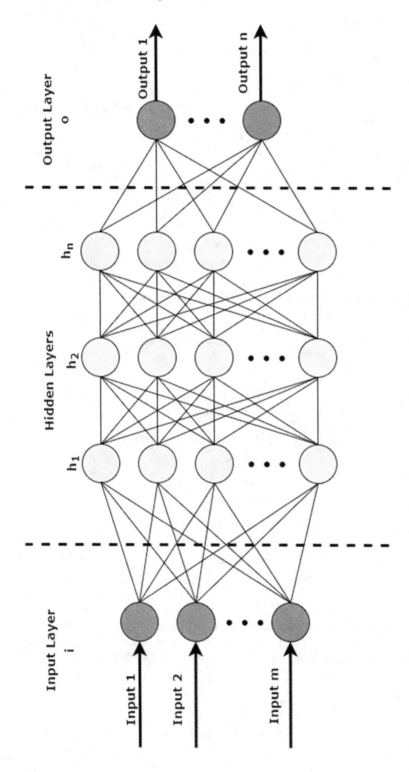

FIGURE 5.3 Architecture of Deep Neural Network.

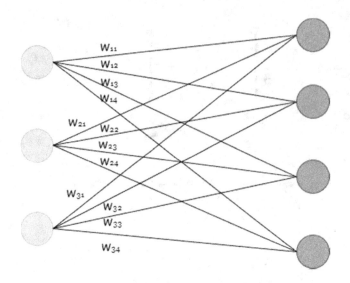

FIGURE 5.4 Connection Between Two Fully Connected Layers.

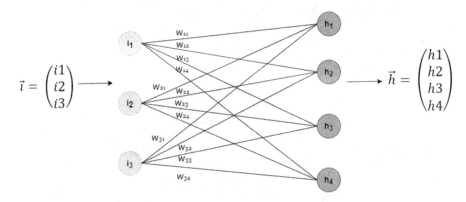

FIGURE 5.5 Neural Network Connection.

$$\vec{h} = \sigma(\vec{z})\qquad\qquad(5.3)$$

The above equation will be called an equation of forward propagation. In order to attain the final prediction vector h, activation function σ is useful to the vector z. The above procedure is recurrent for the complete network to get the output vector o.

$$\vec{h}_1 = \sigma\left(\vec{i}^{\,T} \cdot w_1\right)\qquad\qquad(5.4)$$

$$\vec{h_2} = \sigma(\vec{h_1} \cdot w_2) \tag{5.5}$$

$$\vec{h_3} = \sigma(\vec{h_2} \cdot w_3) \tag{5.6}$$

$$\vec{O} = \sigma(\vec{h_n} \cdot w_{n+1}) \tag{5.7}$$

The activation function is used to familiarize non-linearities in the neural network and it performs non-linear plotting from z to h. Non-linearities in a neural network help us to approximate complex functions efficiently [7]. The sigmoid, hyperbolic tangent (tanh) with a Rectified Linear Unit (ReLU) are the most commonly used activation functions.

5.2.2.1 Threshold Function

The threshold function is a Boolean function. This function passes on 0 when the summed-up assessment of the given input extents a definite threshold and it passes on 1 if the summed-up value is equal to or more than zero. The threshold representation is depicted in Figure 5.6.

5.2.2.2 Sigmoid Function

The sigmoid function provides much smoother output when compared to threshold function and it has gradual progression from 0 to 1. In the sigmoid function, a minor change in input value causes only a small variation in output in contrast to the threshold function. The sigmoid function is cast off mainly for linear regression. The activation functions of sigmoid are depicted and calculated in Figure 5.7 and Equation 5.8.

$$y_i = \frac{1}{1 + \exp(-x_i)} \tag{5.8}$$

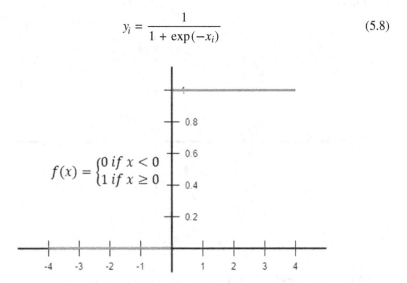

$$f(x) = \begin{cases} 0 \text{ if } x < 0 \\ 1 \text{ if } x \geq 0 \end{cases}$$

FIGURE 5.6 Threshold Function.

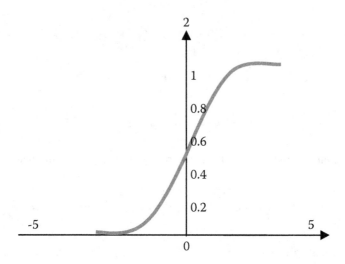

FIGURE 5.7   Function of Sigmoid Activation.

5.2.2.3   Hyperbolic Tangent Function

The hyperbolic tangent function is comparable to the sigmoid function. The only variance is that in a sigmoid function, the value may be anywhere between 0 and 1, whereas the value in hyperbolic tangent function assortments ranges from –1 to 1. The sigmoid function neural network might get stuck during the training phase when there is a portion of negative input, which retains the output nearby to zero. This muddles the learning process while using the sigmoid function, whereas the hyperbolic tangent function gives improved consequences when training neural networks. The activation functions of tanh are depicted and calculated in Figure 5.8 and Equation 5.9.

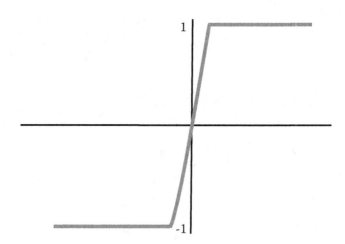

FIGURE 5.8   Hyperbolic Tangent Activation Function.

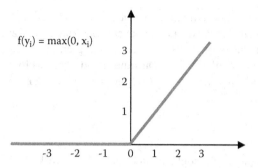

$f(y_i) = \max(0, x_i)$

FIGURE 5.9 ReLU Activation Function.

$$y_i = \tanh(x_i) \qquad (5.9)$$

5.2.2.4 Rectifier Function

The rectified linear unit or in short ReLU is the most popular and most efficient activation function in the world of neural networks. The rectifier function is gradual and smooth after the sharp rise at zero. This means that output could be "no" or some percentage of "yes." The rectifier function does not necessitate any complex calculations and normalization. The activation functions of ReLU are depicted in Figure 5.9.

The second step in the learning process is to compare the prediction vector with the ground truth table ô, whereas the output vector $\vec{o}$ contains the prediction value calculated by the neural network and the vector ô contains the actual values.

The difference between prediction vector $\vec{o}$ and the actual value vector ô can be calculated mathematically by describing the loss function that is contingent on the variance between these values. The higher difference between these two values tends to a higher loss value and the lower difference tends to a lower loss value.

The loss value has to be minimized to increase the accuracy of prediction by a neural network. The minimizing loss function directly makes the neural network do improved calculations. The loss functions used to minimize the loss value are loss functions based on cross-entropy and mean squared error.

5.3 DEEP LEARNING FRAMEWORKS

This section lists the most prevalent open-source deep learning frameworks, including TensorFlow, Keras, caffe, Torch, deeplearning4j, Microsoft Cognitive Toolkit and Theano.

5.3.1 TENSORFLOW

TensorFlow [8] is the most powerful deep learning tool that is open source. It was created by Google in 2015. The tools comprise APIs for Go, Java, Python and C++ that are explicitly premeditated for deep learning tasks and designed for proficient computation of data flow graphs. It provides a high-level API for various neural network layers, namely convolutional layers, pooling layers and fully connected layers and also includes methods for incorporating an activation function and implementing dropout regularization. On other hand, TensorFlow Lite is a machine learning framework that affords an Android neural networks API for mobile and embedded devices.

5.3.2 KERAS

Keras is a Python-based sophisticated neural network API that works with TensorFlow, Theano and CNTK [9]. The Keras asset is that it may run on mutually CPUs and GPUs and supports both convolutional and recurrent networks. Convolutional layers, pooling layers, dropout, regularization and batch normalization are all assisted utility layers.

5.3.3 THEANO

Theano [10] is a Python library that allows accomplishment of fast mathematical computation on multidimensional arrays on both central processing units (CPUs) and graphical processing units (GPUs). Furthermore, it stayed premeditated to outperform further libraries like NumPy and Scikit in terms of accomplishment rapidity and symbolic graph computation. Theano was created in 2008 by the University of Montreal's Montreal Institute aimed at Learning Algorithms and is not extensively sustained since announcement 1.0.0. Theano is proficient in code optimizations that enable it to run at its full potential on the hardware available.

5.3.4 TORCH

Torch is a systematic computing device that offers data in multiple dimensions and logical operations on it, as well as data structures for machine learning algorithms [11]. It's a scripting language that's simple to use, dependable and fast, with a C/CUDA implementation at its heart. Torch is based on the representation of a line that is dynamic. It consents the operator to make improvements to the computational graph as it is being implemented. The Torch Foundation has recently launched PyTorch, and its ease of use has increased its popularity.

5.3.5 CNTK (COMPUTATIONAL NETWORK TOOL KIT)

The Cognitive Toolkit is a free and open-source tool for modelling neural networks using a directed graph and a sequence of computation steps. Microsoft Corporation developed it, and it's correspondingly acknowledged as CNTK (Computational Network Toolkit) [12]. This could be used in a Python or C++ programme as a library, or as a impartial machine learning framework using the Brain Script language. Users can use the CNTK to cascade well-known models such as recurrent networks, convolutional networks, and deep neural networks.

5.3.6 DEEPLEARNING4J

Skymind has been developing and implementing Deeplearning4j, since 2014, a, open source commercial, robust, and distributed deep learning platform has been available in Scala and Java [13]. DL4j can import models commencing additional distributed deep learning frameworks like as Theano, Caffe and TensorFlow when used in combination with Hadoop, Apache and Spark. It's also a domain-specific, high-performance programming language for configuring deep neural networks. It's a domain-specific, high-performance language for configuring multi-layer deep neural networks (DNNs). DL4j bridges the gap among data scientists and business developers, rendering deep learning deployment in big data applications more straightforward. It is compatible with CPUs plus GPUs.

5.3.7 CAFFE

Berkeley at the University of California created Caffe, a well-known deep learning system [14]. It's free and open source, and it's licensed under the BSD license. Caffe is inscribed in C++ libraries and has Python and MATLB command line interfaces. It also supports CNNs, RCNNs, RNNs, DBNs and feeds fully linked neural networks, as well as other deep learning models. Together, CPUs and GPU-based acceleration computational libraries like as Intel MKL, NVIDIA and cuDNN are supported by Caffe. Caffe's benefit is castoff in a variety of academic research projects as well as engineering applications like video, voice and vision.

5.4 CLASSIFICATION

Scene classification is measure of the artificial neural network–based deep learning methods. Deep learning replicas can solve a variety of problems. Deep learning models can be alienated into two categories: scene classification and object detections. The scene classification is mainly used to classify the scenes by using various scene classification models: convolutional neural network (CNN) [15] and deep CNN models [16]. Similarly, the intention of object detection is to classify the scene and additionally localize the object by means of a drawing bounding box over the scene.

5.4.1 CONVOLUTIONAL NEURAL NETWORK (CNN)

A convolutional neural network (CNN) is one of the feed-forward neural networks that can be cast off for image classification, hand written recognition, pattern recognition, etc. The general manner of a convolutional network is revealed in Figure 5.10. The CNN model comprises several components that can be mentioned as follows:

- Convolution layer
- Activation layer
- Sub-sampling or pooling layer
- Flatten layer and fully connected layer
- Soft-max classifier

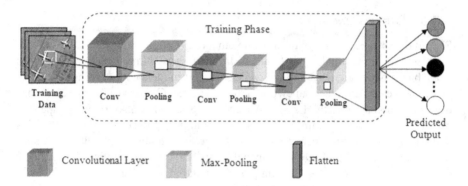

FIGURE 5.10   Architecture of Convolutional Neural Network.

5.4.1.1   Convolutional Layer

A convolutional layer is a major part of the CNN and it is also the origin of such a convolutional neural network. The convolutional process can be defined as follows:

- Define the three-dimensional matrix of the input image (h × w × c)
- Add the padding values to the input matrix
- Place the kernel matrix into the original input matrix
- To perform sliding operation over the input until it reaches the right bottom of the matrix

The aim of the convolutional layer is to excerpt the low-, middle- and high-level features from the input images. The convolution process is calculated as given (Figure 5.11).

As shown in Figure 5.11, for the instance when N = 6 and k = 3, there are four unique places commencing left to right and four unique places starting top to bottom that the kernel could yield.

5.4.1.2   Activation Function

The use of an activation function in the neural network enables complex functional mapping between the inputs and outputs of a neuron to be learned. There are dissimilar kinds of activation functions that exist: sigmoid, tanh and ReLU.

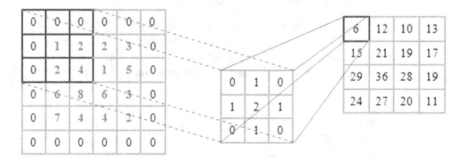

FIGURE 5.11   Convoluted Process in CNN Model.

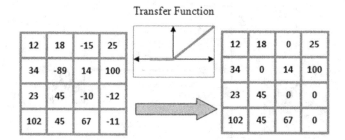

FIGURE 5.12   Activation Function in Pictorial Representation.

Sigmoid:

Range lies to between zero to one (0 to 1)

Tanh:

Range lies between –one to one (-1 to 1)

5.4.1.3   ReLU

ReLU is solitary of the important activation functions in the CNN. The followings steps are mentioned for the process of ReLU operation:

- Every negative value is replaced with zero
- Range lies between zero to infinity (0 to ∞)

Figure 5.12 illustrates the graph representation of ReLU activation function. It can be calculated as mentioned in Equation (5.10):

$$y_i = \begin{cases} 0 & \text{if } x_i < 0; \\ x_i & \text{othereise.} \end{cases} \qquad (5.10)$$

5.4.1.4   Pooling Layer

The determination of pooling layer is to condense the feature resolution of the convolved images. There are three kinds of pooling layers available, such as average pooling, max pooling and min pooling.

5.4.1.5   Max Pooling

The max pooling finds the largest value in the given window. The following steps are for the max pooling operation:

- Define pooling size (generally pooling window size 2 × 2)
- Define stride length

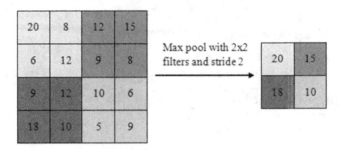

FIGURE 5.13 Max Pooling Operations.

- Move the window across the feature map
- Pick the largest values from each window

The example of max pooling operations was mentioned in Figure 5.13. In this case, we have the pooling window size w = (2, 2) and the size of stride lengths s = (2, 2). It means, in this case the size of the features will be reduced to half of the size. The input matrix for the pooling layer is 4 × 4 and the corresponding output of the pooling values is 2 × 2.

5.4.1.6   Min Pooling

The min pooling finds the lowest value in the given window. The following steps are for the min pooling operation:

- Define pooling size (generally pooling window size 2 × 2)
- Define stride length
- Move the window across the feature map
- Pick the smallest values from each window

The example of min pooling operations was mentioned in Figure 5.14. In this case, we have the pooling window size w = (2, 2) and the size of stride lengths s = (2, 2). It means, in this case, the size of the features will be reduced to half of the size. The input matrix for the pooling layer is 4 × 4 and the corresponding output of the pooling values is 2 × 2.

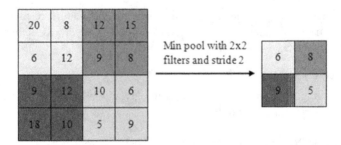

FIGURE 5.14 Min Pooling Operations.

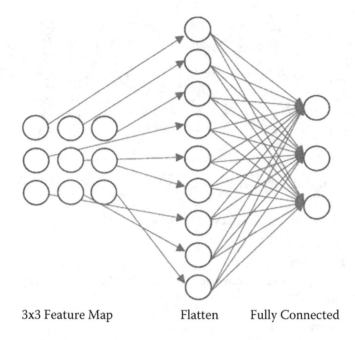

3x3 Feature Map   Flatten   Fully Connected

FIGURE 5.15 Flatten and Fully Connected Layers.

5.4.1.7 Fully Connected Layer

At the end of every CNN model we have to use at least one fully connected layer. The following steps are for the fully connected layer operation:

- Flatten the feature maps (convert the 2D data into 1D data)
- Compute the score for each class

As shown in Figure 5.15, a 3 × 3 feature map is converted into one-dimensional data and nine neurons are fully connected to three neurons (three class outputs).

5.4.2 ALEX NET MODEL

The AlexNet model was developed by Krizhevsky et al. in 2012 [17]. It is one of the first deep convolutional neural networks (deep CNNs) and also the backbone for the many deep CNN models. The architecture of a standard deep CNN (AlexNet) prototype is presented in Figure 5.16. The model consists of three levels that can be summarized as follows:

- Consists of 8 layers
- Five convolutional layers + 5 pooling layers
- Three fully connected layers (2 FC and 1 output classifier)

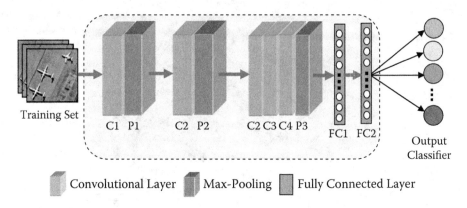

Training Set
C1 P1
C2 P2
C2 C3 C4 P3
FC1 FC2
Output Classifier

▮ Convolutional Layer ▮ Max-Pooling ▮ Fully Connected Layer

FIGURE 5.16 Architecture of AlexNet Model.

5.4.2.1 First Level

The first convolutional layer convolved with an 11 × 11 receptive field (kernel size) and with a tread of 4.

5.4.2.2 Second Level

The second convolutional layer convolved with a 5 × 5 amenable field and with a tread of 4.

5.4.2.3 Third Level

The next three convolutional layers are convolved with a 4 × 4 amenable field and with a tread of 4. The next two layers (sixth and seventh) are fully connected layers with 4,096 neurons. The final layer (eighth) is soft-max, which is used to classify the output.

5.4.3 VGG-16 Model

The VGG-16 model was developed by Simonyan et al. in 2015 [18]. The model consists of five blocks that can be summarized as follows:

- First two blocks have two convolutional layers
- Rest of the three blocks have three convolutional layers
- Total of 13 convolutional layers
- Five pooling layers
- Two fully connected layers
- One soft-max classifier
- The convolutional layer size is progressively improved from 64 to 512

The design of the VGG-16 model is shown in Figure 5.17.

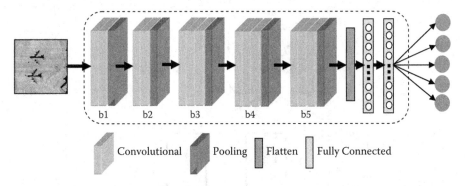

FIGURE 5.17   Architecture of VGG-16 Net Model.

5.5 OBJECT DETECTION

In this segment, we describe the different types of object detection algorithms in deep learning models. In general, object detection would be categorized into two types: region-based detection (two-stage detection approaches) and regression-based detection (one-stage detection approaches).

Region-based convolutional neural networks (R-CNNs) are a deep learning model proposed by Ross Girshick et al. for computer vision and in particular for object detection [19]. R-CNN uses discriminatory exploration to extract 2,000 regions of the given image and these regions are called region proposals. This process cut off the delinquent of choosing a huge number of regions for classification by reducing the number of regions to 2,000. A selective search algorithm is cast off to generate region proposals.

The image data in generated region proposals are then converted into a form compatible with the architecture of CNN by placing them in tight bounding squares of required size. The converted images (277 × 277 pixel size) are then passed through five convolutional layers and two fully connected layers, which act as a feature extractor and produce a feature vector of dimension 4,096 as output. The output layer holds the feature extracted as the image and these features are then passed to the SVM algorithm in order to forecast the manifestation of objects inside the region proposals and to classify them. The working principles of region-based CNN is shown in Figure 5.18.

Drawbacks of R-CNN:
- Classifying 2,000 regions per image drastically increases the training time of the network.
- On average, R-CNN takes nearly 47 seconds for every image and hence it cannot be applied for real-time applications.
- The fixed nature of the selective search algorithm may lead to the compeer's bad region suggestions as no learning process is involved during that phase.

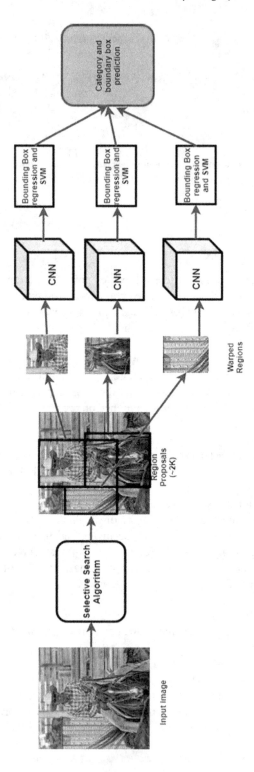

FIGURE 5.18 Working Principle of Region-Based CNN Model.

5.5.1 Fast R-CNN

The improvised version of R-CNN is fast R-CNN, in which the input image is passed to the underlying CNN only once, whereas in R-CNN 2,000 forward passes through CNN happen for a single image [20]. Consider if we have to train the network with the data set of 2,000 images; we may end up with 4M (2,000 × 2,000) forward passes, which is huge and would take lot of time, storage and computational power.

Fast R-CNN consists of CNN and it acts as its backbone. Here, in fast R-CNN, the final pooling layer of CNN is replaced with ROI pooling and the absolute. The FC layer is replaced by a category soft-max layer and a bounding box regression for each category.

The whole input image is nourished into CNN and the features are extracted. The scope of the output feature map is determined by the backbone CNN used. In parallel to feature extraction, the region proposal algorithm-like selective search is used to obtain object proposal windows. The portion of the feature map that fits to this window is then passed into the region of interest (ROI) pooling layer. The operational principles of fast R-CNN are depicted in Figure 5.19.

The ROI pooling layer is a modified description of a spatial pyramid pooling layer with one pyramid level. The main goal of the ROI layer is to reshape arbitrary-sized inputs into a fixed-length output in order to satisfy the size constraint in fully connected layers. This layer divides selected region proposal windows of size h × w obsessed by a H × W grid of sub-windows of size h/H × w/W and a pooling operation is performed on each of these sub-windows. This leads to the immovable length output feature irrespective of the input size.

The fixed length of features obtained from the ROI pooling layer is passed into FC layers, soft-max layers and category-specific bounding box regression layers. The soft-max classification layer is used to generate probability values for each ROI and to catch all background categories. The category-specific bounding box regression layer is used to create more precise bounding boxes.

5.5.2 Faster R-CNN

The faster R-CNN uses a region proposal network (RPN) to engender the region suggestions instead of using a CPU-based region proposal algorithm like a selective search algorithm. Usage of RPN reduces the region proposal generation time from 2 s to 10 ms and also leads to overall improvement in feature representation by allowing the region proposal layer to share layers with the detection layers.

Fast R-CNN is made of two units [21]. The first unit is a deep convolutional network, which recommends regions and then the second unit is the fast R-CNN detector, which practices the projected regions. The working philosophies of faster R-CNN are depicted in Figure 5.20.

A region proposal network [22] is cast off to yield suggestions for object detection by learning from feature maps obtained from a base network such as VGG16, ResNet, etc. The steps in RPN include generation of anchor boxes, classification of anchor box into the foreground and background and fitting the anchor

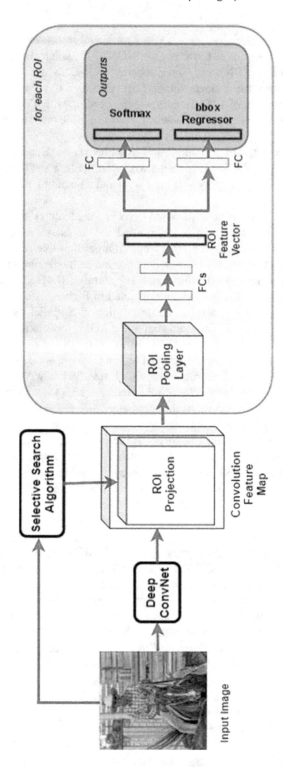

FIGURE 5.19 Working Principle of Fast R-CNN Model.

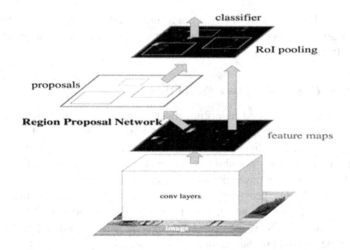

FIGURE 5.20 Working Principle of Faster R-CNN Model.

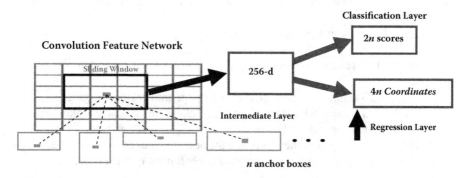

FIGURE 5.21 Working Principle of Region Proposal Network.

boxes to the objects by learning shape offsets. Figure 5.21 depicts the working principles of a region proposal network.

The RPN works in the following way:

1. The feature map engendered through the last convolutional layer in the shared convolution network is considered and a trifling network is slid above the feature map.
2. The output of this is then served into two fully connected network layers: classification layer and regression layer.
3. The red box in Figure 5.21 is called an anchor box. Every anchor box is associated with an aspect ratio and scale.
4. The regression layer creates the coordinates for each of maximum "n" box as output (4n outputs).
5. The classification layer creates the probability that each of the "n" boxes contains an object or not as output (2n outputs).

The RPN can be trained by following an "image-centric" sampling strategy and backpropagation and stochastic gradient descent were used to train the method from beginning to end (SGD).

5.5.3 YOU ONLY LOOK ONCE (YOLO)

YOLO is a regression-based algorithm mostly used in real-time object detection where we need large improvements in speed with a small trade-off in accuracy [23,24]. In YOLO, a distinct convolutional network is used to prophesy several bounding boxes and class possibilities for certain boxes all at the same time.

YOLO has numerous paybacks over contemporary approaches of object detection. First, in YOLO, detection is considered as regression problematic, which avoids a complex pipeline and makes it extremely fast. Second, YOLO is considered a quick R-CNN; it uses the entire picture while making predictions, resulting in less background errors. Third, since YOLO learns generalized representation artifacts, it outperforms top object detection methods including RPN and deformable component models when trained on original images and evaluated on art (DPM).

The YOLO works in the following way:

1. The YOLO framework initially applies a distinct neural network to the full image.
2. The YOLO framework then divides the image into regions or grids.
3. The YOLO framework then applies to each field; image classification and localization are performed.
4. The YOLO framework then forecasts the bounding boxes for the entities in the image and also computes the class probabilities of each object.

YOLO network is divided into two major units: feature extractor with feature detector (multi-scale detector). The image is passed into a feature extractor whenever extracting features of the image and then it is passed to a multi-scale feature detector, which generates bounding boxes for detected objects.

YOLOv3 uses DarkNet-53 and residual network (ResNet) for feature extraction to make feature detection followed by convolutional layers. Darknet-53 is the network pre-trained on ImageNet with 53 convolution layers. The DarkNet-53 network is erected with residual blocks; each block is composed of 1×1 and 3×3 convolutional layers braced with a shortcut connection. The final feature map has 1/32 slighter spatial resolution than the given input image.

The first layer of YOLO has a grid resolution 1/32 of the input image and it detects huge objects in the input image. The last layer of YOLO has a grid resolution 1/8 of the input image and it detects smaller objects in the input image. Between the YOLO layers there are several up-sampling layers and convolutional layers; each layer is composed of a leaky ReLU activation function, sub-convolutional layers and batch normalization. The shortcut connections in a YOLO network are used to connect the intermediate layers of DarkNet-53 with the layers after the up-sampling network. The architecture of the YOLOv3 model is shown in Figure 5.22.

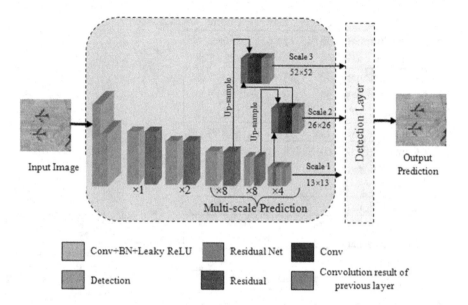

FIGURE 5.22 Architecture of YOLO Model.

The 53 layers of the Darknet are then added with 53 added layers on behalf of the detection head, which makes YOLOv3 a fully connected deep convolutional network with 106 layers. The YOLOv3 is an improvised version with 53 convolutional layers for feature extraction, enhanced detection metric and logistic regression-based predicting technique for class probabilities of each bounding box and uses binary cross-entropy loss-based class predictions during training. The multi-scale prediction of the YOLO model is shown in Figure 5.23.

5.6 SINGLE SHOT DETECTOR

The single shot detector (SSD) is the single-shot multi-object detecting model, unlike RPN-based approaches that perform object detection in two shots: one for engendering region proposals and another for distinguishing the object from each region proposal [25]. This makes SSD much faster than RPN-based approaches.

A SSD makes use of feed-forward convolution network to produce collection bounding boxes of fixed size and confidence score for the occurrence of object class in those bounding boxes and this step is followed by final detections using non-maximum suppression.

The base network of SSD is composed of customary architecture used for image classification and the auxiliary structure is added to the setup to perform detection. The significant features of auxiliary structure are as follows:

Multi-scale feature maps: Convolutional feature layers are auxiliary to the completion of the base network. These layers provide recognition at multiple scales.

Convolutional predictors: Each feature layer added or the one as a base network yield a set of calculations for detection by making use of convolutional filters.

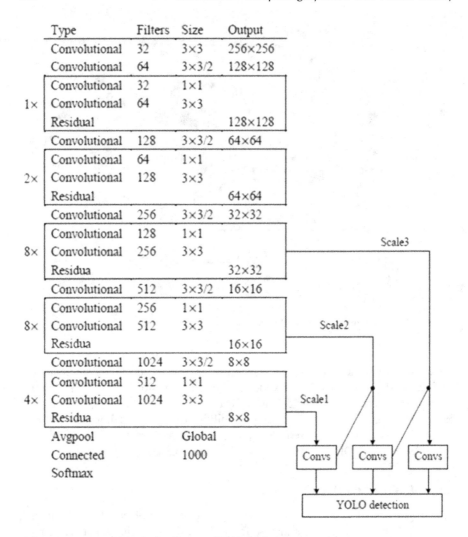

Type	Filters	Size	Output
Convolutional	32	3×3	256×256
Convolutional	64	3×3/2	128×128
1× Convolutional	32	1×1	
Convolutional	64	3×3	
Residual			128×128
Convolutional	128	3×3/2	64×64
2× Convolutional	64	1×1	
Convolutional	128	3×3	
Residual			64×64
Convolutional	256	3×3/2	32×32
8× Convolutional	128	1×1	
Convolutional	256	3×3	
Residua			32×32
Convolutional	512	3×3/2	16×16
8× Convolutional	256	1×1	
Convolutional	512	3×3	
Residua			16×16
Convolutional	1024	3×3/2	8×8
4× Convolutional	512	1×1	
Convolutional	1024	3×3	
Residua			8×8
Avgpool		Global	
Connected		1000	
Softmax			

FIGURE 5.23   Multi-Scale Prediction of YOLO Model.

Default boxes and aspect ratios: For multiple feature maps, each feature map cell has a collection of default bounding boxes. The offsets value relative to the defaulting bounding box shapes in each feature map cell is expected, as thriving as the trust score indicating the existence of a determined class instance respectively of those boxes.

As shown in the Figure 5.24, SSD's architecture is constructed over the VGG-16 architecture, whereas the fully connected layers of VGG-16 are discarded. VGG-16 achieves remarkable performance in high-quality image classification tasks, which makes it the base layer of SSD. Fully connected layers of VGG are traded with conventional auxiliary convolutional layers, thus making it possible to remove features at various scales and the size of each subsequent layer's input is decreased progressively.

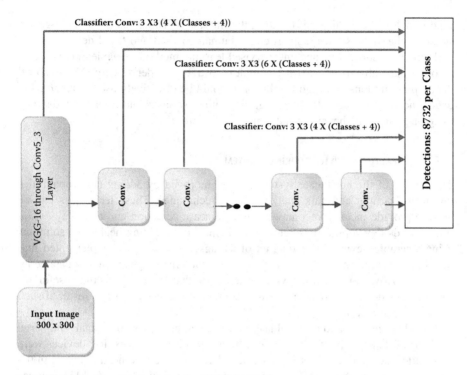

FIGURE 5.24 The Architecture of a SSD Model.

5.7 APPLICATIONS OF MULTIMEDIA ANALYTICS BASED ON DEEP LEARNING TECHNIQUES

The key presentations of deep learning models are discussed in this segment. Deep learning models offer a better method for dealing with big data analysis because they can handle vast volumes of labelled and unlabelled data. Sentiment analysis of social media data, multimedia summarization, digital forensics, data journalism, natural sciences, text processing, natural language processing and urban computing are only a few of the areas where the deep learning model has been used. Deep learning models derive high-level sorts from these input data and are imperative to attain categorized depictions of the features. Deep learning devices concerned a growing number of high-tech developers, including Microsoft, Facebook, Twitter and Google. To solve complex problems, deep learning models like CNN, deep CNN, RNN and LSTM are cast off.

5.7.1 BIG DATA ANALYTICS

Big data is an evolving trend in which increasingly large amounts of data are collected in various fields and valuable information is extracted for decision making [26]. Traditional methods for extracting data from massive volumes of data are not efficient. As a result, deep learning is critical in big data analytics and is commonly used in fields like medical informatics, the Internet of Things, national intelligence, social media, cybersecurity, smart cities, human conduct and fraud detection.

Using both supervised and unsupervised learning methods, deep learning models automatically absorb and extract complex abstractions derived from hierarchical data representations at a high level [27]. It could be effectively used to resolve big data concerns like semantic indexing, discriminative tasks and semantic video and image tagging, whichever are not conceivable with predictable practices.

5.7.2 HEALTHCARE MONITORING SYSTEM

The quality of the healthcare system can be improved by adding new capabilities like deep learning and data analytics-based techniques. These techniques can be used to provide medical recommendations, disease detection and also play a vital role towards developing personalized medicines and building healthcare solutions using wearable devices. The Internet of Things (IoT) has been implemented into healthcare systems in recent years. This has resulted in a large amount of data being produced. However, handling vast volumes of data while extracting useful information as healthcare databases has become more complicated as a result of such rapid data production.

Deep learning techniques would support the healthcare system in analyzing large amounts of data and predicting disease. In critical circumstances, IoT devices were cast off to track patient illnesses and guide alerts to those who need to know. In the healthcare sector, about 65% of the Internet of Medical Things (IoMT) policies have been cast off [28]. The Internet of Medical Things (IoMT) helps to change the healthcare industry by enabling the transition from disorganized to synchronized treatment. In addition, IoT sensors are cast off to monitor human day-to-day health behaviors including walking, biking, cycling remoteness measurement and snooze analysis, among others [29].

Asthana et al. [30] created a human healthcare method that recommends wearable devices to everyone. This device gathers all of the user's health information, such as medical records and current IoT data from sensors or hospitals. The algorithm then uses deep learning classification techniques such as convolutional neural networks besides recurrent neural networks to make predictions about human health diseases. When it comes to analyzing data, medical insurance providers often use deep learning techniques. Human diseases are also predicted and detected by the healthcare system at an early level.

5.7.3 SOCIAL MEDIA

For the large number of Internet users, social media sites have developed into an important fragment of their everyday lives. People use social media to increase their erudition. Apart from knowledge, people can use social media to showcase their talents by creating content. A video résumé, for example, is something that professionals make and post on social media to demonstrate their presence. Images, text, emotions and videos are all examples of content types. Users create an enormous amount of data that exhibit all of the characteristics of big data since there are few restrictions on content production on social media. This knowledge can be used in a number of company analytical and predictive applications. The development in multimedia analytics plays a vivid role in obtaining data related to disaster from several sources like sensors, social media and videos to make rescuing, searching and disaster site inspection more efficient than ever.

5.8 CONCLUSIONS

Huge volume of multimedia data are being engendered worldwide every day; in the last decade multimedia analytics has gained much research interest and had a drastic growth with the proliferation in storage capacity, computation authority and the advancement of new state-of-the-art techniques like machine learning and deep learning. Machine learning and deep learning models can be built and trained using huge volumes of available data. These models are then used to make decisions. It provides an overview of multimedia big data, deep learning and numerous frameworks to work with deep learning models. Deep learning models perform classification and object detection in multimedia data and the applications of multimedia analytics in various fields. So, here we explained various up-to-the-minute deep learning models for classification like CNN, AlexNet, VGG-16 and object detection like fast and faster R-CNN, YOLO and SSD. Moreover, we have discussed applications of multimedia analytics and many deep learning contexts. The intention of this chapter is to persuade researchers to explore big data analytics using deep learning and apply it for unraveling different research complications.

REFERENCES

[1] S. Vrochidis (2019). *Big Data Analytics for Large-Scale Multimedia Search*, Hoboken, NJ: John Wiley and Sons Ltd.
[2] G. Cavallaro, M. Riedel, and M. Richerzhagen (2013). On Understanding Big Data Impacts in Deep Learning Models, *IEEE Journal of Applied Earth Observations and Remote Sensing*, vol. 8, no. 10, pp. 1–13.
[3] M. Najafabadi, F. Villanustre, T. Khoshgoftaar, N., Seliya, R. Wald, and E. Muharemagic (2015). Deep Learning Applications and Challenges in Big Data analytics, *Journal of Big Data*, vol. 2, https://doi.org/10.1186/s40537-014-0007-7.
[4] P. Deepan and L. R. Sudha (2021). Deep Learning and its Applications related to IoT and Computer Vision, *Artificial Intelligence and IoT: Smart Convergence for Eco-friendly Topography*, pp. 223–244, https://doi.org/10.1007/978-981-33-6400-4_11.
[5] B. B. Le Cun, J. S. Denker, D. Henderson, R. E. Howard, W. Hubbard, and L. D. Jackel (1990). Handwritten Digit Recognition With a Backpropagation Network, *NeurIPS Proceedings*. Citeseer, Denver, CO.
[6] R. Hecht-Nielsen (1989). *Theory of the Backpropagation Neural Network, inIJCNN*, San Diego, CA.
[7] Andrew L. Maas, Awni Y. Hannun, and Andrew Y. Ng. (2013). Rectifier Nonlinearities Improve Neural Network Acoustic Models. *ICML Workshop on Deep Learning for Audio Speech and Language Processing*, vol. 28. Computer Science Department, Stanford University, CA.

[8] https://www.tensorflow.org/ (2017). *An Open-source Software Library For Machine Intelligence.*

[9] https://keras.io/ (2017). *Keras: The Python Deep Learning Library.*

[10] http://deeplearning.net/software/theano/.

[11] http://torch.ch/ (2017). *Torch: A Scientific Computing Framework For LuaJIT.*

[12] https://docs.microsoft.com/en-us/cognitive-toolkit/ (2017). *The Microsoft Cognitive Toolkit.*

[13] https://deeplearning4j.org/ (2017). *Deep Learning for Java: Open-source, Distributed, Deep Learning Library for the JVM.*

[14] http://caffe.berkeleyvision.org/ (2017). *Caffe.*

[15] Deepan, P. and Sudha, L. R. (2020). Remote Sensing Image Scene Classification using Dilated Convolutional Neural Networks, *International Journal of Emerging Trends in Engineering Research (IJETER)*, vol. 8, no. 7, July 2010, pp. 3622–3630, ISSN: 2347-3983.

[16] Deepan, P. and Sudha, L. R. (2020). Object Classification of Remote Sensing Image Using Deep Convolutional Neural Network, *The Cognitive Approach in Cloud Computing and Internet of Things Technologies for Surveillance Tracking Systems*, pp. 107–120, https://doi.org/10.1016/B978-0-12-816385-6.00008-8.

[17] Alex Krizhevsky and Geoffrey E. Hinton (2014). Image Net Classification with Deep Convolutional Neural Networks, arXiv: 1409.1556, pp. 1–9.

[18] K. Simonyan and A. Zisserman (2014). Very Deep Convolutional Networks for Large-scale Image Recognition. arXiv: 1409.1556.

[19] R. Girshick, J. Donahue, T. Darrell and J. Malik (2016). Region-Based Convolutional Networks for Accurate Object Detection and Segmentation, *IEEE Transactions on Pattern Analysis and Machine Intelligence*, vol. 38, no. 1, pp. 142–158, https://doi.org/10.1109/tpami.2015.2437384.

[20] R. Girshick (2015). Fast R-CNN, 2015 *IEEE International Conference on Computer Vision (ICCV)*, Santiago, Chile, pp. 1440–1448, https://doi.org/10.1109/ICCV.2015.169.

[21] S. Ren, K. He, R. Girshick and J. Sun (2017). Faster R-CNN: Towards Real-Time Object Detection with Region Proposal Networks, *IEEE Transactions on Pattern Analysis and Machine Intelligence*, vol. 39, no. 6, pp. 1137–1149, https://doi.org/10.1109/tpami.2016.2577031.

[22] P. Tang, X. Wang, A. Wang, Y. Yan and W. Liu (2018). Weakly Supervised Region Proposal Network and Object Detection, *Conference in ECCV*, pp. 1–19.

[23] J. Redmon, S. Divvala, R. Girshick and A. Farhadi (2016). You Only Look Once: Unified, Real-Time Object Detection, 2016 *IEEE Conference on Computer Vision and Pattern Recognition (CVPR)*, Las Vegas, NV, USA, pp. 779–788, https://doi.org/10.1109/CVPR.2016.91.

[24] P. Deepan and L. R. Sudha (2021). Effective Utilization of YOLOv3 Model for Aircraft Detection in Remotely Sensed Images, *Materials Today: Proceedings*, Elsevier, ISSN: 2313-4534, https://doi.org/10.1016/j.matpr.2021.02.831.

[25] W. Liu et al. (2016). SSD: Single Shot Multi-Box Detector, *Computer Vision – ECCV*, pp. 21–37.

[26] X. Chen, and X. Lin (2014). Big Data Deep Learning: Challenges and Perspectives. *Access, IEEE*, vol. 2, pp. 514–525, https://doi.org/10.1109/ACCESS.2014.2325029, 2014.

[27] P. Deepan and L. R. Sudha (2019). Fusion of Deep Learning Models for Improving Classification Accuracy of Remote Sensing Images, *Journal of Mechanics of Continua and Mathematical Sciences (JMCMS)*, vol. 14, pp. 189–201, ISSN: 2454-7190.

[28] J. PandiaRajan and S. Edward Rajan (2019). Smart-Monitor: Patient Monitoring System for IoT Based Healthcare System Using Deep Learning, *IETE Journal of Research*, pp. 1–9, https://doi.org/10.1080/03772063.2019.1649215

[29] Xiao Ma, Zie Wang, Sheng Zhou, Haoyu Wen, and Yin Zhang (2018). Intelligent Healthcare Systems Assisted by Data Analytics and Mobile Computing, *Wireless Communications and Mobile Computing*, vol. 12(1), pp. 1–16.

[30] S. Asthana, A. Megahed and R. Strong (2017). A Recommendation System for Proactive Health Monitoring Using IoT and Wearable Technologies, *IEEE*, Honolulu, pp. 14–21, https://doi.org/10.1109/AIMS.2017.11

6 An Optimal System on Data Challenge with Distributed Data Management on Cloud, Fog and Edge Computing

M. Arvindhan[1], Abhineet Anand[2] and Md. Abdul Wassey[2]
[1]Galgotias University, Greater Noida, India
[2]Chitkara University Institute of Engineering and Technology, Chitkara University, Punjab, India

6.1 INTRODUCTION

The needs of business change quickly. Increasing the quantity and variety of information generated in digitalization, the international Internet of Things (IoT) and social media are creating a data-driven community. This pattern also contributes to new consumer requirements and the current proliferation of digital disruptors. Traditional data centers are being set on fire against this backdrop. Many current systems are static and cannot integrate various types of data. In addition, ad-hoc reporting is also not supported and forward-looking questions cannot be answered. There is a greater need for scalable, agile and efficient data storage solutions than ever before. A new data platform is about to be adopted. Officials of all shapes and sizes – public and private alike – repeatedly complain about the lack of access to correct, timely, appropriate and reliable information. This is necessary to encourage decision making. However, the road from data collection to actionable insight is often blocked or at least partly blocked. Interestingly, a data server is for most of those who complain. Even after the specifications were met, the information demands continue to evolve, and are now overworked desperately [1].

Traditional warehouses of data are as rigid as concrete. Various types of data (structured and unstructured) cannot be collected in many organizations. They have time-consuming reports, data quality problems and, most significantly, cannot respond to forward-looking, predictive questions. It should be open science to open up all research fields. Open access to research data will help speed up discovery by allowing reuse and reducing duplications to provide increased value for financed research. Open

DOI: 10.1201/9781003196686-6

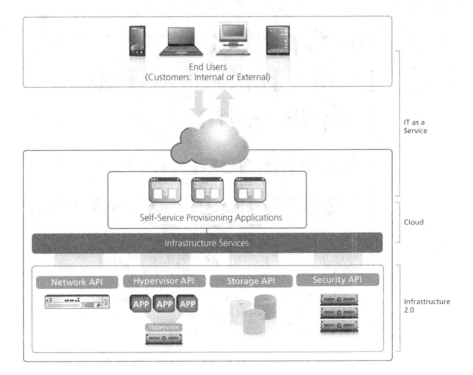

FIGURE 6.1   Architecture Model for Cloud Connecting Structure.

data, including unexpected new findings, and economic gain, are demonstrated to make the studies more efficient, more likely to be quoted and open up innovation to the good of society (Figure 6.1). Based on observed attitudes, researchers need not be persuaded of these more benefits. Numerous studies have shown a good acknowledgement of the advantages of data sharing and high degree of willingness to share and use data.

The complex foundation of infrastructure services and the automation and orchestration architecture is needed for IT as a service. Without such a basis, customers can quickly have the necessary calculation or storage resources for the application, but cannot specify the services required for security, access control, availability and performance.

This includes an integration with the supply and management networks and a business infrastructure. The lack of standards across an infrastructure spectrum pose a challenge for service enabling and, in some cases, organizations do not have an API to achieve service enabling [2,3].

6.2 INFRASTRUCTURE AND INTEGRATION WITH PROVISIONING AND MANAGEMENT SYSTEMS

6.2.1 SERVICE-ENABLING THE INFRASTRUCTURE

To automate and, eventually, organize operational activities and processes, the infrastructure is essential to enable service. Enabling services is a challenge because

FIGURE 6.2 Analytics Structure of API Methods.

the infrastructure demesne does not have the required standardization. APIs and SDKs are also very granular and are unique to the technologies of the infrastructure parts (Figure 6.2). Different API calls may be required for common operating tasks, each infrastructure component requiring different calls with their own specific terminology. For example, creating a VLAN may require a switch to make very different service calls from the load balancer. Not only product-specific expertise is needed for these discrepancies, but strong knowledge of production tools and processes and networking [4].

6.2.2 THE GOAL OF AUTOMATION AND ORCHESTRATE METHODS

Automation and orchestration are frequently combined, but two terms are distinct. Automation seems to be the coding of a task, including "adding this server to the loading balancing pool" or "dirigating web requests to a different data center." Orchestration is a method codification, including such as "deploying an application," normally involving several automated tasks.

Automation and orchestration are vital for the performance and the scope of data center operations, as well as supporting IT as a service. Task automation makes it possible to create repeatable processes – orchestration – which can lead to more efficiencies by simplifying processes for implementation and maintenance [5].

6.2.3 INTEGRATION OF CONFIGURATION MANAGEMENT AND IMPROVISATION ENGINES

To achieve true elasticity, many components in the data center must be orchestrated. Providing or decommissioning another instance of an application is just the first phase in a far more complex process involving infrastructure-wide load balancing, acceleration and optimisation, protection and networking components. The triggers for provisioning and decommissioning of applications are equally critical for elasticity and automated deployment. The above mechanisms typically apply thresholds to performance and availability set by businesses and operational requirements and need indicators to measure those thresholds. Visibility of metrics is not the only important thing. The mechanisms by which such metrics can be transmitted should also be made possible, such as triggers and alignment with reporting systems. Leading supply and orchestration motors enable companies to take advantage of these integrations immediately as the basis for the implementation of private clouds [6].

BIG-IP system integrates with: Automation and Orchestration

a. VMware vCloud Director and vSphere methods
b. IBM Pure Systems
c. HP Cloud Mappings
d. Microsoft System Center 2012
e. CloudStacks
f. Puppet Labs Puppet
g. Opscode Chef

6.2.4 CODIFYING DEPLOYMENT POLICIES

The incorporation of renewable policies is essential if the aim is to provide IT as a provision, move to a hybrid architecture, or organizational consistency. A application management plan which would accept and enforce appropriate policies governing protection, efficiency and availability, allows organizations, regardless of their environment, to achieve greater economies of scale within operations and maintain consistent application placements.

Such policies, however, should be flexible to ensure that criteria unique to locations and applications can be implemented on an individual or project basis. This means taking feedback in such a way that the regulation is abstracted without sacrificing its precise implementation inside the configuration [7].

Solution for Codifying deployment:Transition to a Hybrid Model

However, companies that are happening the road to a hybrid model need an architectural solution that really can support inter-cloud requirements. In hybrid architectures, having the ability to link environments is critical, and organizations need to consider managing their identity with access in a multi-cloud setting.

Hybrid architectures include distributed infrastructure implementation and computer services, including safety and availability including access management policies. If this disjoint policy set was deployed, operational anomalies in the delivery of applications can lead to inconsistent availability and performance and fail to meet appropriate business and operational requirements [8].

6.2.5 Inter-cloud Architecture

It can be intimidating to develop a cloud platform to move to a hybrid architecture. In order to meet such inter-cloud requirements and standards in later times, hybrid architectures can require major changes to the data center architecture. This calls for stable interconnectivity between the private and public cloud environments and anticipates how systems span environments and also the infrastructure elements to be replicated in a hybrid model's public cloud (Figure 6.3).

By adding BIG-IP WAN Optimization Manager (WOM) WAN optimization features, traffic around securely interconnected sites is optimized. The optimum performance enables businesses to migrate virtual machines live and to keep business stakeholders' performance standards [9].

The varying or ignored policies can affect efficiency, safety and availability. The lack of functionality in cloud environments or an operator's failure during the deployment phase can hinder monitoring and visibility. The multiple cloud designs of private clouds attract applications that may suffer from the same administrative inconsistencies, although this may be more likely in an inter-cloud setting.

6.2.6 Addressing Topological Dependencies

Because operations are used by one approach to handle local resources, and cloud-deployed resources by alternative, processes and policies disconnect and disintegrate.

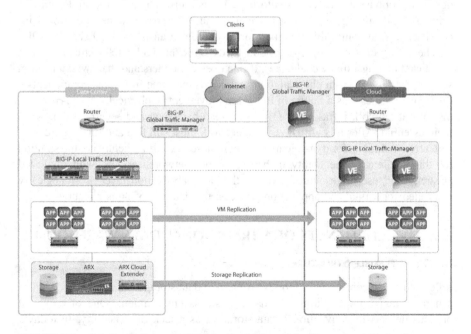

FIGURE 6.3   Solution Allows for Bridging and Virtualizing Inter-cloud Technology.

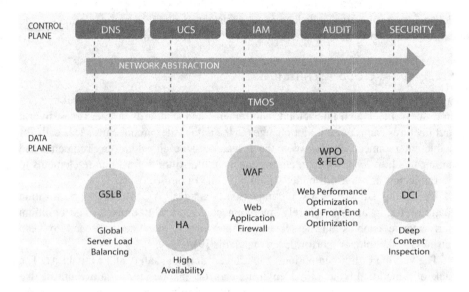

FIGURE 6.4 Platform Isolates Control and Data Plane Responsibilities, Enabling Organizations to Virtualize the Network by Abstracting Services.

This type of inconsistency management contracting and organizational risk management costs more.

The inconsistent or ignored policies can affect efficiency, safety and availability. The lack of functionality in cloud environments or an operator's failure during the deployment phase can hinder monitoring and visibility. The multitenant design of a private cloud attracts applications that may suffer from the same operational inconsistencies, although this may be more common in an inter-cloud setting [10].

The effect of switching from static to dynamic network architectures is most significant amongst many challenges. If one does not understand that switching from static to dynamical, from physical to virtual, needs the same functional components – the same facilities – as a failure of the entire movement in physical world. As part of the overall network infrastructure, firewall, load balancing and security services remain essential. After it is virtualized, the network must still live and run (Figure 6.4).

Current solutions for dynamic management in server infrastructure are the solution for this challenge. The ability to handle virtual network services with improved elasticity and use of resources involves the same abstraction layer as strategically controlled for virtualized application services: the delivery of an application level.

6.3 FIVE CHALLENGES OF A TRADITIONAL DATA WAREHOUSE

6.3.1 INFLEXIBLE STRUCTURE

Inflexibility is a common but serious problem in data storage. The lack of versatility can have major implications in today's business environment. One of the most obvious defects in conventional data storage was a lack of versatility. In today's

FIGURE 6.5 Technology for Traditional Data Warehousing.

volatile market environment, this is a specific problem. Mergers and acquisitions are at large, and a community where knowledge is accessible on demand has been generated by the app economy and consequent IT consumption. IT architecture must be flexible and adaptable, allowing decisions and regular changes to move forward (Figure 6.5). A simple request to modify a data model can take months with an inflexible data warehouse, involve many individuals and require entirely new data sources. It takes businesses a lot of time, money and commitment [11].

6.3.2 COMPLEX ARCHITECTURE

There is a shortage of inclusion. Multiple technologies forming a complicated structure also lack a native process integration. As a result, ownership costs are raised, governance problems have been resolved and mobility has been lost. Companies with complex architecture in data warehouses lack access to simple, practical insights without a single source of reality. This prevents them from making educated choices. Many instruments have the same or at least very similar functions. This means spending money in duplicate, unprofitable innovations [12].

6.3.3 SLOW PERFORMANCE

The increase in data volumes in the source system. Traditional warehouses will stop unprecedented amount of data. The increase in volumes of extracts. More data means migrating additional information, which, for conventional data warehouses, can prove complex. Outdated processes for reading and writing. The process of preparation is drawn from shifting data between slow disks. Methods that are ineffective. Many conventional data stores replicate and do not reuse data, complicating the process of planning [13].

6.3.4 OUTDATED TECHNOLOGY

Processors: More frequent and substantial updates are required for outdated CPUs to meet new requirements. The challenge is to have a processor in an integrated server as a single element. Therefore, reform without updating the economic platform is difficult or even impossible. Remember. Antiquated servers also mean that the central data hub's output is impeded by slow memory processing. Stocking. Many conventional data warehouses use basic hard drive arrays that fight to fulfill the demands of increased user inquiries.

Networking: – Building a network. Old principles of networking and suboptimal routing from data storage, origin and business intelligence networks lead to bottlenecks and deletions [14].

6.3.5 LACK OF GOVERNANCE

Systems source: When businesses propose improvements to their sources, conventional data storage will complicate the study of impacts. It can also make mapping and cataloging new systems difficult without violating the rules on data management.

Extracting, changing, loading: In a typical warehouse, ETL processes cannot generate standard log data, making it harder to display and search. They may also lack consistent procedures, resources and controls to ensure the proper processing of sensitive data. Furthermore, trying to load new data sets quickly in an obsolete warehouse can undermine mechanisms for data governance [15].

Store and optimize: Traditional data storage facilities also do not have structured data models or rich metadata for semanthropic exploration. Furthermore, it does not support common rules for data segmentation (Figure 6.6).

6.4 THE IMPORTANCE OF ADAPTATION

6.4.1 FLEXIBILITY

Enterprises need to be fast, agile and able to make quick and well-informed decisions more and more. These characteristics include a high degree of versatility data platform. With virtual data access, SAP HANA ensures this versatility, which reduces the need of aggregate data. In addition, regardless of its source, type or structure, the platform can store and analyze data.

6.4.2 CONSOLIDATE IT INFRASTRUCTURE

Companies want streamlined architecture to minimize overall ownership costs and tackle governance problems. The overall complexity of SAP HANA is reduced in

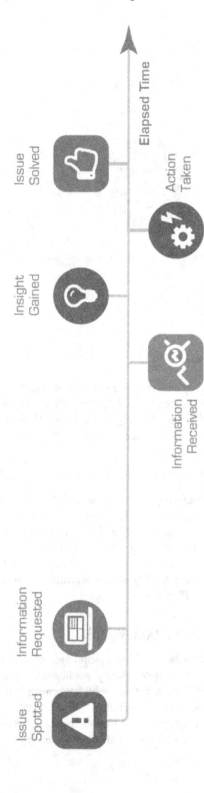

FIGURE 6.6 Information Received and Flow of Elapsed Time.

FIGURE 6.7   Modern Model of Data Control and Governance.

several respects. First, which unites data into one structure so that the local analyses can minimize the movement of the data. Second, their integration of the memory processing power of SAP HANA and the large capacity of Hadoop removes the need for modeling and indexing data in order to improve performance [16].

6.4.3 SCALABILITY

Some of the SAP tools enrich high-speed analysis and transactions during the duration of data processing. Its memory architecture drops the neck of the disk bottle and significantly accelerates output in a fraction of a second to provide correct answers. At about the same time, the virtual modeling of the platform enables in-situ analysis and reduces data movement and eliminates activities such as aggregation and replication. Its compatibility with powerful integration tools makes data supply and loading quicker (Figure 6.7).

6.5   EDGE TAXONOMY AND FRAMEWORK

Basically, edge computing is distributed cloud computing, which includes many network interconnected application modules. Many applications today are already distributed: (1) a smartphone application with a cloud backend; (2) a consumer device, such as thermocommunication or a voice control device that connects directly to the cloud; (3) a smart clock or sensor connected to a smartphone, and then the cloud. Moreover, many network features, including dedicated privately owned networks,

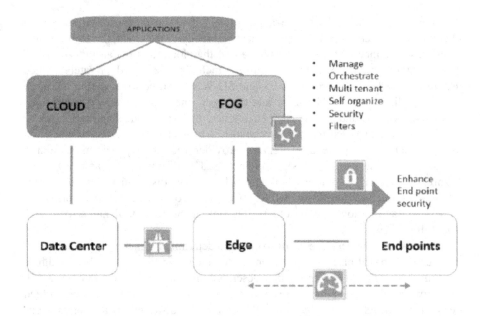

FIGURE 6.8   Application Distribution towards Data Center on Different Domain.

fixed wireless connectivity and SD-WAN and network slicing are more and more distributed to the public and thus meet the requirements of many businesses and vertical industries (Figure 6.8). They enable new business models and use cases [17].

In order to achieve a balance between different business lenses (i.e., cloud, telecom, cable, IT, OT/industrial, consuming) and high-level taxonomy categories, based on major technological and logistic agreements, edge computing and the related taxonomic terms introduced in this document were carefully developed. These bargainings include whether a computer resource is able to accommodate program abstraction (e.g., through containers and/or virtual machines), whether it is available or in a physically protected data centre, and whether it is on a LAN or WAN related to process/use (an important consideration if a use case is latency-critical vs. -sensitive). Without using the terminologies of the edge, this paper aims to provide a holistic view of the fact that telecom operators generally use the words "near" or "far" edges to differentiate between the user-close infrastructure and the infrastructure further upstream (near edge). The fact that the relative position is seen in the eyes of the provider rather than the consumer can be confusing. In another example, in some circles, the terms "thin" and "thick" were used to define degrees of on-site computing capacity, but these terms do not match up between user edge resources that are physically protected at a data centre, as opposed to those that are distributed at accessible locations [18].

6.5.1   Architectural Trends at the Service Provider Edge

Basically, edge computing is distributed cloud computing, which includes many network-integrated application modules. Many of the current applications, such as

(1) a cloud-backend smartphone application, (2) a consumer device such as a thermostat or voice control system connected directly to the cloud, (3) a smartwatch or a sensor connected to the smartphone and the cloud, and (4) an industrial IoT (IIoT). Moreover, many network features, including dedicated privately owned networks, fixed wireless connectivity and SD-WAN and network slicing are more and more distributed to the public and thus meet the requirements of many businesses and vertical industries. They enable new business models and use cases.

The volume of PoPs inside your network is projected as well as the extension of their networks to the edge service provider for cloud providers including Amazon Web Services (AWS), Google Cloud Platform (GCP) and Microsoft Azure. Each cloud provider would probably try to distinguish the edge offer of their products in one way, while others will concentrate on AI workloads. Users can supply tools, while others can look at their IoT toolchains with edge capabilities [19].

The service provider edge needs to ensure a deterministic method to measure and enforce QoE based on key applications needs including latency and bandwidth in accordance with the design principles described previously. As most Internet traffic is encrypted, these guarantees are probably focused on the transport layer, which leads to the development algorithms that assess the distribution rate for the congestion control. For regional data isolation policies for stores and workloads, a similar concept-10 theory will develop beyond compliance with global legislation on data security (Figure 6.9).

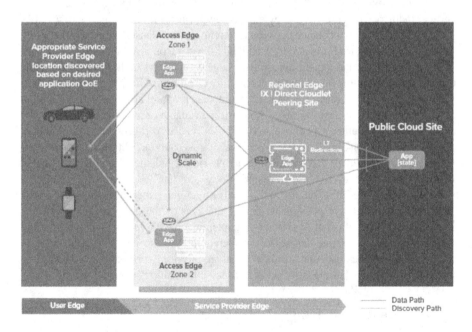

FIGURE 6.9   Service Provider on Edge Connection with Cloud Site.

6.5.2 SAFETY AND MANAGEMENT OF DISTRIBUTED EQUIPMENT

Constrained and intelligent device edges are normally used in semi-secure positions in the field to be readily accessible. As such, a null-confidence security model is important and it is not necessary that a system should be behind a network firewall. The distributed computer resources in all cases require a remote software upgrade to avoid expensive truck rolls and develop over time through scalable, software-defined architecture in the event of on-site data centers and smart devices. IoT and smart device edge client-centered computer resources are able to use MANO devices that embrace containerization and virtualization in abstract, and have security headrooms such as data encryption. Limited devices, meanwhile, use embedded software images that are normally suited to the host hardware and need to use a more capable system for additional security measures immediately afterwards [20].

Big data typically shares some or all of these features with development sources:

1. Generated digitally – i.e., data is produced digitally and can be processed using series of numbers or zeros (as opposed to being digitalized manually), so that computers can be manipulated [21].
2. Produced passively – by daily product or by interaction with digital services.
3. Automatic collection – i.e., the system is in place which extracts, stores and generates the relevant data.
4. Trackable geographically or by time – e.g., details on the location of cell phones or call time.
5. Constant analysis – that is, knowledge is important and can be analyzed in real time to human well-being and growth.

It is important to note that "real time" does not necessarily mean happening immediately for the purposes of global growth. "Real time" can instead be interpreted as information generated and made available in a relatively short and appropriate time frame, and details made available in time frames that enable the actions required in answer to be taken, i.e., to create a feedback loop. It is mainly the intrinsic time dimensionality of the data that defines its characteristics jointly as actual time and those of the feedback loop. It can also be added that the real-time nature of the data depends on the analysis carried out in real time and, if necessary, in real time [21].

6.5.3 EDGE COMPUTING USE CASES

In the case of large amounts of capital like centralized cloud data centres, where the service providers and user borders consequently use models to provide inferences on the local level, the general tendency is to create profound learning and model training at the edge of the artificial intelligence and machine learning (AI/ML) (Figure 6.10). The positioning and performance of models along the periphery depends on a number of factors, such as latency issues being tackled, autonomy ensured, bandwidth usage reduced, end-user privacy improved and data sovereignty requirements [22–24].

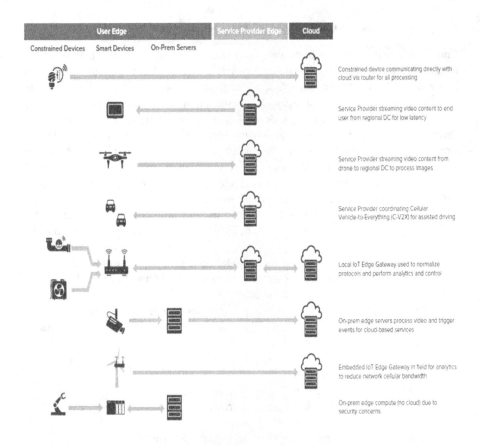

FIGURE 6.10   On Perm Edge Computing to Cloud Security.

a. Industrial with IoT
b. Computer Vision
c. Augmented Reality (AI)
d. Virtual Reality (VI)
e. Gaming
f. Unmanned aerial vehicles

6.6   THE FOG COMPUTING DATA PLATFORM

While Cisco first coined the word fog computing, other parties have investigated and created similar concepts. Three of these technologies are detailed below, including some of their main distinctions from fog systems. For edge computing, a more comprehensive comparison can be found in. (1) Edge controllers (PACs), which handle data processing, storage and communications, perform localized processing on a system. It offers an advantage over fog computing because it decreases failure points and makes every computer more autonomous (Figure 6.11).

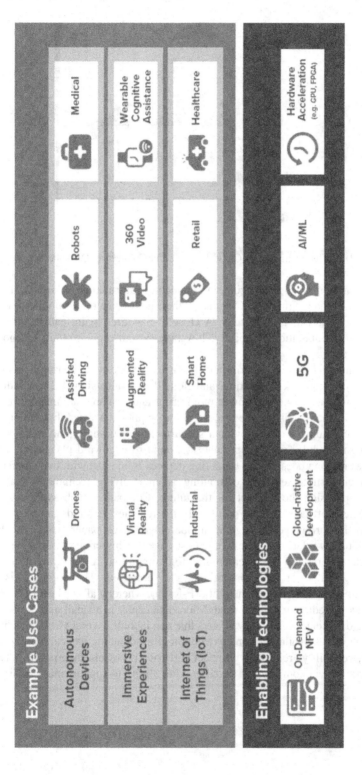

FIGURE 6.11 Different Use Case Model on Enabling Technology.

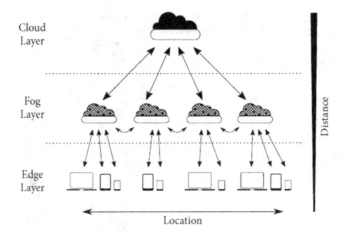

FIGURE 6.12   Different Layers of Cloud location and Connecting Distance.

The same function makes managing and collecting information in large-scale networks like IoT challenging (2). Cloudlet is the middle part of a "mobile device – cloud – cloud" three-stage hierarchy. A cloudlet has three main features: its fully auto-managed computational capacity, low latency and regular cloud technology [25] (Figure 6.12).

Through using iFog-Sim, performance measurements can be measured and edge devices, sensors, network connections and cloud data centers simulated. In addition, iFog-Sim combines replicated power monitoring and resource administration services at two distinct levels: device placement and application scheduling. In order to accommodate a number of deployment scenarios: a) cloud-only placement, where all application modules operate in datacenters and b) edgeward placement, where fog modules run on nodes next to bordering devices is bundled with two application module placing strategies. Fog computing is an evolving technology that will certainly help design the Internet for the future. Companies such as Cisco, Dell, Intel and Microsoft take a stance to identify the relevant standards in an open-fire consortium. In current years, a great number of research papers have also been published, but few relate to the field of data storage. Each section presents the state-of-the-art data storage of the research area fog computing. An instantly decodable binary code method of data dissemination is proposed. The approach is implemented on a fog radio access network (F-RAN), whereby all files are distributed over devices by reducing contact time. Microdata center is a small and completely functioning data center with many servers that can include several virtual machines. Microdata center is available to many technologies, including fog computing, such as latency, reliability, relatively compact, security protocols incorporated, bandwidth utilization via compression saved and the ability to support many additional services [26,27].

6.6.1 MANAGING RESOURCES IN MICRODATA CENTRES

In addition to allowing advanced technology, fog computing can accomplish numerous systems level tasks including the management, prediction, evaluation and reservation of computing resources. It can also filter data on policy basis, preprocessing and improve safety measures. Nebula frameworks used for the management of other systems' computer resources are very likely to share problems in technology (discussed in the section "Defined and Virtualized Software for Radio Access Networks"). Another important hazard is the one of the malicious insider who can breach user-to-user, user-to-user, admin-to-user and administrator level access control [27,29].

6.6.2 DISTRIBUTED DATA STORAGE MECHANISMS

Control and data storage is carried out centrally in fog computing. However, it is not realistic to store anything in the cloud, due to (a) high latency, (b) high demand for traffic between the border and cloud, and (c) high storage costs. Furthermore, the mobility of fog or edge devices is considered a critical parameter. The aforementioned disadvantages could be resolved through the distribution of stored data on the fog layer according to different criteria, such as the geographical position of producers and consumers [30,31].

6.6.3 DATA DISTRIBUTION

Data sharing techniques may remove data storage constraints by distributing the data equally to all nodes in large networks with a large number of cooperative devices. Data can go to the fog nodes in fog computing. To our awareness, fog computing contains a number of relevant works for this purpose In addition, in several research activities over the last few years, P2P storage systems have been proposed as a distributed storage solution. Meanwhile, in special systems such as content delivery systems (CDN) and cloud computing P2P storage techniques have been used [32,33].

6.6.4 DATA DISSEMINATION

Data disclosure is a well-researched topic in WSNs. In addition, diffusion algorithms are used in the fastest distance in low bandwidth networks. Dissemination algorithms may also be adjusted to determine various transmission routes based on the current network connections load. Dissemination mechanisms between different fog nodes can be used in fog computing to minimize overall traffic by choosing the best method. The dissemination of data occurs mainly between fog nodes (Table 6.1).

TABLE 6.1
Potential Security Issues Found in Fog Applications

Attack category	Possible threats	Possible solutions	Impact
Virtualization issues	Hypervisor attacks VM-based attacks Weak or no Logical Segregation Side channel attacks Privilege Escalation Service abuse Privilege escalation attacks Inefficient resource policies	Multi-factor Authentication Intrusion Detection System User data isolation Attribute/identity based encryption Role-Based Access Control model User-based permissions model Process isolation	As all services and VMs are executing in a virtualized environment, its compromise will have adverse effect on all Fog services, data and users
Web security issues	SQL injection Cross-site scripting Cross-site request forgery Session/Account hijacking Insecure direct object references Malicious redirections Drive-by attacks	Secure code Find and patch vulnerabilities Regular software updates Periodic auditing Firewall Anti-virus protection Intrusion Prevention System	Exposure of sensitive information, attacker can become legitimate part of network, and enable malicious applications to install
Internal/external communication issues	Man-in-the-Middle attack Inefficient rules/policies Poor access control Session/Account hijacking Insecure APIs and services Application vulnerabilities Single-point of failure	Encrypted communication Mutual/Multi-factor authentication Partial encryption Isolating compromised nodes Certificate pinning Limiting number of connections Transport layer security (TLS)	Attacker can acquire sensitive information by eavesdropping and get access to unauthorized Fog resources
Data security related issues	Data replication and sharing Data altering and erasing attacks Illegal data access Data ownership issues Low attack tolerance Malicious Insiders Multi-tenancy issues Denial of Service attacks	Policy enforcement Security inside design architecture Encryption Secure key management Obfuscation Data Masking Data classification Network monitoring	High probability of illegal file and database access, where attacker can compromise both user and Fog system's data
Wireless security issues	Active impersonation Message replay attacks Message distortion issues Data loss Data breach Sniffing attacks Illegal resource consumption	Authentication Encrypted communication Key management service Secure routing Private network Wireless security protocols	Vulnerable wireless access points can compromise communication privacy, consistency, accuracy, availability and trustworthiness
Malware protection	Virus Trojans Worms Ransomware Spyware Rootkits Performance reduction	Anti-malware programs Intrusion Detection System Rigorous data backups Patching vulnerabilities System restore points	Malware infected nodes will lower the performance of the entire Fog platform, allow back-doors to the system and corrupt/damage data permanently

6.6.5 PRIVACY PRESERVATION IN FOG COMPUTING

The framework also provides the ability to reduce the number of functions in order to minimize data transmissions to fog nodes. The purpose of this work is to preserve personal and sensitive data when transmitting them. Choosing a crypt and key management algorithm will boost the proposed methodology, which focuses on those who play a major role in data privacy maintenance (Figure 6.13). Moreover, there is little discussion about the computer overheads necessary for comprehensive data manipulation before and after contact (fuzzing, segregation, encryption, decryption, ordering, reordering). Fog-related technologies are also explored, including edge computing and cloudlets. Most fog applications have been found not to regard protection as part of the framework, but to concentrate on features that makes many fog platforms vulnerable. Furthermore, literature describes the broad range and potential of fog

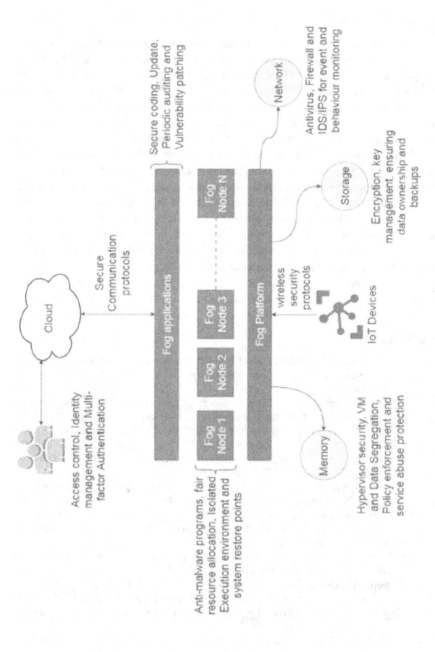

FIGURE 6.13 Fog Computing Technology and Protection Solutions Removing Different Security Defects.

computing applications that all require a high degree of protection to protect consumer data from the CIA [34,35].

6.7 CONCLUSION

One of the benefits of using public cloud computing is that the network infrastructure abstraction has already been accomplished, which already addresses network planning problems for cloud computing. As companies progress with their own private cloud projects, they often face the same or related obstacles, and for this reason they create their own cloud computing environment. Although these priorities can vary in organisations, over time, the network challenges are similar. Understanding the problems and their solutions would enable companies to identify a less risky path to success. In particular, a versatile yet stable framework on which cloud models can be founded is crucial for the readiness of the network in terms of service enhancement, integrating and automation. A decision-making support framework of this kind will entail the development of high-quality information, established security risks and the solution of best practices that can be formalized as a predictive system or regulations, policies and facts. In view of the current application scenario and available information, the framework will also need to include and describe an inferenze engine that could provide a suitable solution or advice. A fog network is linked to both end users and cloud platforms, along with massive data processing, storage and transmission, using only minimal resources.

REFERENCES

[1] I Stojmenovic, S Wen (2014). The fog computing paradigm: Scenarios and security issues. In: *Computer Science and Information Systems (FedCSIS), 2014 Federated Conference*. IEEE. pp 1–8.

[2] JY Kim, H Schulzrinne (2013). Cloud support for latency-sensitive telephony applications. In: *Cloud Computing Technology and Science (CloudCom), 2013 IEEE 5th International Conference On*, vol. 1. IEEE. pp 421–426.

[3] F Bonomi, R Milito, Zhu J, Addepalli S (2012). Fog computing and its role in the internet of things. In: *Proceedings of the First Edition of the MCC Workshop on Mobile Cloud Computing*. ACM. pp 13–16.

[4] P Sareen, P Kumar (2016). The fog computing paradigm. *International Journal of Emerging Trends in Engineering Research* 4:55–60.

[5] LM Vaquero, L Rodero-Merino (2014). Finding your way in the fog: Towards a comprehensive definition of fog computing. *ACM SIGCOMM Computer Communication Review* 44(5):27–32.

[6] K Saharan, A Kumar (2015). Fog in comparison to cloud: A survey. *International Journal of Computers and Applications* 122(3):10–12.

[7] Dastjerdi AV, Gupta H, Calheiros RN, Ghosh SK, Buyya R (2016) Fog computing: Principals, architectures, and applications. arXiv preprint arXiv:1601.02752.

[8] Mahmud R, Buyya R (2016). Fog computing: A taxonomy, survey and future directions. arXiv preprint arXiv:1611.05539.

[9] Cisco (2015). Cisco fog computing solutions: Unleash the power of the internet of things. Online: https://www.cisco.com/c/dam/en_us/solutions/trends/iot/docs/computing-solutions.pdf. Accessed 13 Dec 2016.

[10] M Schumacher, E Fernandez-Buglioni, D Hybertson, F Buschmann, P Sommerlad (2013). *Security Patterns: Integrating security and systems engineering.* Wiley.

[11] M Satyanarayanan (2015). A brief history of cloud offload: A personal journey from odyssey through cyber foraging to cloudlets. *GetMobile: Mobile Computing and Communications Review* 18(4):19–23.

[12] Zissis D, Lekkas D (2012). Addressing cloud computing security issues. *Future Generation Computer Systems* 28(3):583–592.

[13] Alliance CS (2016). The treacherous 12 cloud computing top threats in 2016. Online: https://downloads.cloudsecurityalliance.org/assets/research/top-threats/ Treacherous-12_Cloud-Computing_Top-Threats.pdf. Accessed 22 Dec 2016.

[14] I Stojmenovic, S Wen, X Huang, H Luan (2015). An overview of fog computing and its security issues. *Concurrency and Computation: Practice and Experience.* Online. https://doi.org/10.1002/cpe.3485.

[15] S Yi, Z Qin, Q Li (2015). Security and privacy issues of fog computing: A survey. In: *International Conference on Wireless Algorithms, Systems, and Applications.* Springer. pp 685–695.

[16] M. Arvindhan, A. Anand (2019). Scheming an proficient auto scaling technique for minimizing response time in load balancing on Amazon AWS Cloud. In: *International Conference on Advances in Engineering (ICAESMT)*, Uttaranchal University, Dehradun, India.

[17] A Ahmed, E Ahmed (2016). A survey on mobile edge computing. In: *Intelligent Systems and Control (ISCO), 2016 10th International Conference.* IEEE. pp 1–8.

[18] RM Pierson (2016). How does fog computing differ from edge computing? Online: https://readwrite.com/2016/08/05/fogcomputing-different-edge-computing-pl1/. Accessed 12 June 2017.

[19] Li Y, Wang W (2013). The unheralded power of cloudlet computing in the vicinity of mobile devices. In: *Globecom Workshops (GC Wkshps).* IEEE. pp. 4994–4999.

[20] L. Chai, et al., (1999). An Adaptive Estimator for Registration in Augmented Reality. Proc. 2nd Int'l Workshop Augmented Reality (IWAR 99). IEEE CS Press, Los Alamitos, California, pp. 23–32.

[21] S. Yi, Z. Qin, Q. Li (2015). Security and privacy issues of fog computing: A survey. In: *Proceedings of the International Conference on Wireless Algorithms, Systems, and Applications*, pp. 685–695, Springer.

[22] W. Ramirez, X. Masip-Bruin, E. Marin-Tordera et al. (2017). Evaluating the benefits of combined and continuous fog-to-cloud architectures, *Computer Communications* 113:43–52.

[23] O. Salman, I. Elhajj, A. Kayssi, A. Chehab (2015). An architecture for the Internet of Things with decentralized data and centralized control. In *Proceedings of the 2015 IEEE/ACS 12th International Conference of Computer Systems and Applications (AICCSA)*, pp. 1–8, Marrakech, Morocco, November.

[24] H. Chen, H. Zhang, M. Dong et al. (2017). Efficient and available in-memory KV-Store with hybrid erasure coding and replication. *ACM Transactions on Storage (TOS)* 13(3):1– 30.

[25] Y. Jiang, M. Ma, M. Bennis, F. Zheng, X. You (2018). User preference learning based edge caching for fog-ran, https://arxiv.org/abs/1801.06449.

[26] F. H. Fitzek et al. (2016). On network coded distributed storage: How to repair in a fog of unreliable peers. In *Proceedings of the International Symposium on Wireless Communication Systems (ISWCS)*, pp. 188–193.

[27] A. Anand, A. Chaudhary, M. Arvindhan (2021). The need for virtualization: When and why virtualization took over physical servers, advances in communication and computational. *Advances in Communication and Computational Technology.* pp. 1351–1359.

[29] S. Li, M. A. Maddah-Ali, A. Salman Avestimehr (2017). Coding for distributed fog computing, *IEEE Communications Magazine* 55(4):34–40.

[30] S. Alonso-Monsalve, F. Garcia-Carballeira, A. Calderon (May 2017). Fog computing through public-resource computing and storage. In *Proceedings of the 2nd International Conference on Fog and Mobile Edge Computing, FMEC 2017*, pp. 81–87.

[31] S. Sarkar, S. Misra (2016). Theoretical modelling of fog computing: A green computing paradigm to support IoT applications. *IET Networks* 5(2):23–29.

[32] R. S. Carbajo, C. Mc Goldrick (2017). Decentralised peer-to-peer data dissemination in wireless sensor networks, *Pervasive and Mobile Computing* 40:242–266.

[33] M Arvindhan, B. Prakash Ande (2020). Data mining approach and security over Ddos attacks. *ICTACT Journal on Soft Computing* 10:2. https://doi.org/10.21917/ijsc.2020.0292.

[34] M. Arvindhan, Arnav Munshi, Sanchit Sapra (2021). A novel random forest implementation of sentiment analysis. *International Research Journal of Engineering and Technology (IRJET)* 2821–2824.

[35] Abhineet Anand, M Arvindhan (2020). Development and various critical testing operational frameworks in data acquisition for cyber forensics. *Critical Concepts, Standards, and Techniques in Cyber Forensics* 88–102. https://doi.org/10.4018/978-1-7998-1558-7.ch006.

7 Anomaly Detection in Real-Time Videos Using Match Subspace System and Deep Belief Networks

Dr. D. Ratna Kishore[1,2], *Dr. D. Suneetha*[1,3],
Dr. G. S. Pradeep Ghantasala[1,4]
and B. Ravi Sankar[1,2]
[1]Department of Computer Science and Engineering
[2]Andhra Loyola Institute of Engineering and Technology,
 A.P., India
[3]NRI Institute of Technology, A.P., India
[4]Chitkara University Institute of Engineering and
 Technology, Chitkara University, Punjab, India

7.1 INTRODUCTION

Video surveillance systems are used for detecting the anomalies and generating the alerts while monitoring people's activities and generating alerts. Typically, trials of anomalous actions are rare. Despite the importance, accurate determination of anomalies can be very challenging. The scheme is trained to acquire recurrent patterns of ordinary activities from a set of normal samples though use of methods of activity recognition.

7.2 CATEGORIES OF ANOMALY

An imperative facet of the anomaly detection procedure is the flora of the anticipated anomaly. Anomaly is confidential into three types on the basis of information that is shown in Figure 7.1.

7.2.1 POINT ANOMALY

This was the meekest kind of anomaly then is the emphasis of the mainstream of inquires on anomaly detection. In point anomaly, an individual information occasion can be considered as anomalous regarding whatever remains of information [1].

DOI: 10.1201/9781003196686-7

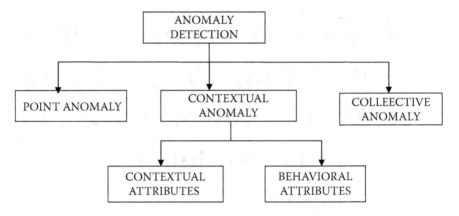

FIGURE 7.1   Categories of Anomaly.

7.2.2   CONTEXTUAL ANOMALY

In contextual anomaly, an information is anomalous in a particular setting (be that as it may, not something else). It is also elevated as a conditional anomaly [1]. The conception of a background is fetched by the edifice in the data set and has to be itemized as a portion of the problem formulation. Respectively, data occurrence is demarcated using the subsequent two sets of qualities:

> (1) Contextual attributes. These contextual attributes are cast off to regulate the context (or locality) for that occurrence. (2) Behavioral attributes. These behavioral attributes describe the non-contextual appearances of an occurrence [2].

7.2.3   COLLECTIVE ANOMALY

If an assortment of correlated information event is inconsistent for the whole information established, it is named a collective anomaly. The personal information examples in a collective anomaly might not be anomalies independent from anyone else, yet, their event together as an accumulation is anomalous.

7.2.4   TECHNIQUES IN ANOMALY DETECTION

Common methods are available for finding anomalies such as digital video tampering detection, spatio-temporal video-volume, and frame deletion detection method, etc. Anomaly detection methods can function in three different modes: semi supervised anomaly detection, and unsupervised anomaly detection and supervised anomaly detection.

7.3   DIGITAL VIDEO TAMPERING DETECTION

Digital video tampering detection can be characterized as passive and active techniques. Passive video altering discovery techniques are grouped into three classifications in the light of the kind of imitation they address: detection of twofold or different compacted recordings, gegion altering location and video between edge falsification identification. What's more, quickly present the preliminaries of video records required for comprehension video-altering phony [3].

7.4 SPATIO-TEMPORAL VIDEO-VOLUME CONFORMATION INSIDE VIDEO

This method is an unsupervised measurable erudition structure in light of the examination of spatial-worldly video-volume design inside the video cubes. It learns worldwide movement examples and neighborhood remarkable conduct designs through bunching and inadequate coding, separately. Upon creation of design word reference gained from typical conduct, a scanty remaking cost measure is intended to notice anomalies that happen in the video together universally and in the neighborhood [4].

7.5 ROTATION-INVARIANT ATTRIBUTE MODELING MOTION COHERENCE (RIMOC)

Rotation-invariant attribute modeling motion coherence highlight is utilized for learning measurable models of typical cognizant movements in a regulated way. A multi-scale conspire connected on a derivation-based strategy permits the occasions with flighty movement to be identified in space and time, as great competitors of forceful occasions [5].

7.6 LEARNING DEEP DEPICTIONS OF ARRIVAL AND MOTION

Arrival and motion deepnet (AMDN) approach in the light of profound neural systems for consequent learning of highlight representations. A novel twofold combination system, integrating the recompences of conventional early amalgamation and late amalgamation procedures for eliminating misuse of the corresponding data of both appearance and movement patterns. In particular, slanted de-noising automatic encoders are proposed for independent erudition of both arrival and movement highlights in addition to a joint representation [6].

7.7 REAL-TIME TROOP COMPORTMENT DETECTION IN VIDEO

Real-time technique is meant for detecting procedures in swarmed video arrangements. This slant depends on the mixture of visual element abstraction and image dissection and everything deprived of the necessity for a preparatory stage. A quantifiable trial assessment had been done on numerous freely accessible video successions, comprising information of different 15 group situations and distinctive sorts of occasions for exhibiting the viability of the approach [7].

7.8   LOCALITY SENSITIVE HASHING FILTERS (LSHF)

Locality sensitive hashing (LSH) capacities sift through unusual exercises. An Internet redesigning technique is likewise brought in to the structure of LSHF for adjusting to the progressions of the visual scenes. Besides, buildup of another assessment capacity is built up for assessing the hash guide and utilize the particle swarm optimization (PSO) technique to hunt down the ideal muddle capacities, enhancing the effectiveness and precision of the anticipated anomaly detection technique [8].

7.9   FRAME DELETION DETECTION METHOD

Detection of abnormal periodical ancient rarities is divide into two elements for gauging the size of variety in expectation lingering then the quantity of intra command blocks. In view of the formulated components, a combined file is focused for catching unusual sudden vagaries in visual streams. A data set is made comprising six subsets, and a trial of the uncovering competence of strategy in mutually GOP (group of pictures) and video levels. The exploratory fallouts demonstrate the steady performance of the frame deletion technique under a different setup [9].

7.10   CHALLENGES

There are numerous challenges in the process of distinguishing human activity after videos, relating to videos occupied with stirring backgrounds, like trees, dissimilar lighting circumstances (day/night time, in/out door); different 16 viewpoints; occlusions; variations within each commotion; hefty number of commotions; and inadequate extents of considered data between others. Some of the challenges are as follows:

- The approaches necessitate a learning passé for estimating countless restrictions of the scheme as the unknown parameter always leads to a false alarm.
- Many detection methods are not adopted for a crowd scene, considering the complexity involved in locating the anomaly present in the video.
- Usually the anomalous observation appears normal, thereby creating the mission of significant typical behavior becoming a tedious process.
- The exact definition of an anomaly is dissimilar for diverse domains. Therefore, application of a method settled in one province to one more is not easy.
- Obtainability of a categorized data set for training/authentication of replicas cast off by anomaly detection approaches is a chief challenge.
- Noise is problematic to differentiate and eradicate from the actual anomalies while often the data contains noise.

One of the most important factors is to provide security and public safety. Nowadays, surveillance videos are in demand, and the demand is increasing day by day [10–13]. So many various applications are present for surveillance videos. One of the concentrated applications is anomaly detections. It is one of the challenges and has substantial consideration from various fields like business and academic. In general, the anomaly occurs at irregular events but rarely for long time videos [14]. The detection of anomalies is one of the challenging tasks because the gathering of all anomalies from a single surveillance video is not possible and it is a desirable solution. To identify the anomalies, one general and common solution is learning problems. With the help of learning the problem, we learn the normal events based upon the training videos and identify and detect anomalies upon the distance of normal events and testing events [15].

Another task for anomaly detection is to extract various features for modeling video events. Many existing studies are present for modeling the videos and identifying various features with respect to different perspectives. In general, to extract low-level features, use a histogram approach [16]. In this approach, calculate a histogram of gradient and also the optical flow to calculate the stint for motion measurements. Some of the existing algorithms that have trajectory-based features are extracts and use the concept of semantic relation between various objects [17].

In this proposed approach, a new novel multimodal representation was introduced to identify and detect anomalies for 3D images, and various complex crowed surveillance videos and images [18]. To identify low-level features of this proposed framework consist of fusing multimodal approach with the help of spatiotemporal energy features. For this one, we proposed a new deep belief networks (DBNs) to achieve our goals. In this first step, the foreground video patches are extracted, and various low-level features are utilized to differentiate subordinate map structures. These subordinate map structures are served as an input for the multimodal to acquire sophisticated features. The one period classifier, i.e., support vector machines [19], are used to identify and detect the anomalies and various anomaly events. The following are the various contributions for the proposed model:

- A new supervised deep learning model is used to earn various middle-level features.
- A new multimodal fusion is introduced to identify high-level features for modeling various crowd events.
- Match subspace system (MSS) also one period Support Vector Machine (SVM) is used to distinguish anomalies in crowd images or videos.
- Compare the presentation of the proposed algorithm through numerous existing algorithms under variously available data sets.

7.11   RELATED WORKS

Various existing algorithms are presented to identify anomalies in 3D images and videos. All of the existing work is broadly classified into two types. Those are trajectory-based methods and spatiotemporal patch-based approaches [20]. In general, trajectory-based methods have different steps [14]. In the first step, we extract various features to model the normal events. Secondly, perform trajectory clustering operation for feature representation to modal the normal events [21].

Yadigar Imamverdiyev et al. [22] projected a novel algorithm for anomaly discovery for 2D and 3D images. In these, similar features are formed as a single class cluster and then apply one class support vector machine to identify and detect anomalies in an image. Keith Hollingsworth et al. [23] presented a new algorithm founded on the interaction energy. Luis Martí et al. [24] projected a new algorithm for addressing dissection problems. Shen Su [25] projected a novel algorithm founded on the correlation dimensions. Adam proposed an approach constructed on the histogram of optical flow regions with the help of a distribution concept. Cy Chen [26] proposed a general social model to identify and analyze the behavior of an image like crowd behavior. Hemank Lamba [27] proposed combinations of various component analyzers to identify and model the events.

Kelathodi Kumaran Santhosh [28] proposed a new algorithm to identify gradient features for 3D images. It uses a high-speed learning framework for model normal/abnormal events. These models detect anomalies with 150 frames per second. The speed of this one is the best one to compare to various existing algorithms. Olga Isupova [29] developed a new hierarchical dynamical texture model of the video, founded on the temporal model and spatial model.

Bin Liu [30] projected a novel algorithm founded on the real-time camera position. In this technique first, excerpt the reduced structures after the input or surveillance image and then identifying the anomaly detection based on the features extraction and motion of the camera. In general, if the motion of the camera was changed, it centrals to alteration in the features of the image. False alarms are obtained due to noise in this algorithm and the noise ratio was gradually less. In some of the techniques, anomaly detection was divided into two different types, i.e., event encoding and anomaly event detection model based on these two models; this model presents different techniques of hidden methods such as Markova models [31]. In order to test the anomaly detection in images, videos and audio, a new type of data set was proposed, i.e., street scene. It consists of more accurate and high pixel images and videos. The noisy percentage of the video was minimum when compared to other data sets.

Amogh Mahapatra [32] proposed a new algorithm based on the context information. It is a mathematical model used for detecting the anomalies of a 3D image. In this we consider the activities that occurred from the space and those are treated as features for extraction. Amit Adam [33] introduced a new procedure for recognition of confident types of infrequent attacks. This procedure is constructed upon the numerous local displays which consist of statistics. Each local monitor raises an alarm whenever there is an unusual attack and those decisions are integrated and finally identify an unusual event. This algorithm requires minimum setup and this is fully automatic to the environment. It was confirmed on variability of real-time packed divisions and results will decrease the false alarm rate. Xuguang Zhang et al. [34] proposed an algorithm for a crowd abnormal detection method using conversion of energy near distribution. This technique not merely reduces the perception effect of the camera, it also detects abnormal detection behavior in time. Finally, in this algorithm, the crowd performance was considered based on the transformation of uniformity, entropy and contract between three descriptors of the co-occurrence matrix. In this paper UMN data set is used for experimental calculation and this shows better results for characterizing anomalies in the videos. Ramin Mehran et al. [35] introduce a new scheme for detecting and localizing abnormal behaviors in crowd videotapes, exhausting the social force model. To achieve this, they placed a grid of subdivisions above the image; besides, it is covered through space time middling with an optical flow. Based on the moving specks, interaction between the navies is predictable, exhausting the proposed social force model. This collaboration force is randomly diagramed into the image level to get the potency course of each pixel in each frame. In this, based on nag of words the frames are classified into abnormal and normal frames. The anomalies are present in the abnormal casings and confined using the collaboration forces and the anomalies present in the normal frames are not localized. There are no anomalies in the normal frames, only those are present in abnormal frames. The experimental results were carried out on the public data sets and the different crowd videos are taken from web. The results proved the proposed approach successfully detects anomalies in abnormal frames and it is purely based on the optical flow.

Li et al. [4] have discussed the problematic of instinctive anomaly detection for reconnaissance applications. In an uncrowded scene, an anomalous event has been detected with a Gaussian mixture model (GMM), particle filtering for adaptive foreground, optimal clustering of features and a statistical scene modeller, which conglomerates trajectory-based as well as region-based information for improved anomaly detection. Venkatesh Saligrama et al. [36] have designated intimate, unsupervised methods to the visual anomaly detection founded on statistical movement analysis of the contextual model. These methodology indications to innumerable anomalies are oscillating since spatially localized plus temporally persistent anomalies, to spatially disseminated anomalies and to structured anomalies in unstructured consequences. In this agenda, action at the respective position is demonstrated by means of a binary state Markov chain that acquires a feature descriptor, like motion vector, size and shape. Moreover, the saliency of signatures, authorizations spatial suggestion inside the camera's arena of vision. This necessitates storing binary values of tumbling memory requests, whereas in certain enactments, it is candid abundant to be measured for entrenched platforms at camera linkage edge.

Yong Wang and Dianhong Wang [37] have proposed a method for detecting forefront regions in visual sequences. For every pixel, the time succession of pixel intensity is deliberated as symbol sequences, which are accomplished and modelled by the Markov prototypical for procurement of the anomaly detection threshold. The earlier founded on obvious occurrence recognition and definition of high-level semantic elucidations of video sequences are two approaches used for all application fields and then based on anomaly detection. Claudio Piciarelli et al. [38] have proposed a concept established on a single-class support vector machine clustering, wherever the originality detection SVM competences are cast off for the empathy of anomalous paths. Various soft computing techniques have been used for identifying the anomalies. Deep learning is a learning method-based algorithm for object recognition. This attempts to model higher-level perceptions in data by consuming a deep graph through numerous dispensation layers. The super resolution (SR) method of traditional span coding is a deep convolution network. Chao Dong et al. [39] have experimented getting high-resolution images from low-resolution images from deep convolution networks. Learning endways mapping has been adopted among low- to high-determination images and this has accomplished qualitative and quantitative results and also produced benefits by giving a solution to a low-level vision problem such as simultaneous SR+ de-noising or image deblurring. It is extended to cope with three contour channels. The deep architecture has the possibility of speedy modeling video-based anomaly detection. It contains single- or multiple-person and activities. Yong Shean Chong and Yong Haur Tay [40] have discussed poignant objects inside the examined location that describes the dynamic-based Bayesian networks and Jianwei Ding et al. [41] have given the tracking-learning data approach for overcoming problems in object tracking. The authors have re-presented and experimented the tracking of the blurred object by a deep image learning approach. Their work mainly focuses on reducing the haziness in videos. Bayesian estimation framework is used for carrying out tracking. The robustness of the severe blur is increased by blurring kernel.

7.12   PROPOSED MODEL

The structure for the overall proposed model is shown in Figure 7.2.

The proposed anomaly event detection algorithm is centered upon the three different stages. First, deep belief networks are familiarized to identify and cram the mid-level features for a crowded image or crowded data. In the second stage, the erudite mid-level feature representation are further combined with multi-model fusion framework to identify high-level representation and obtained the relation among low-level and high-level feature representations. In the last stage, apply the match subspace system (MSS) and the one class support vector machine to

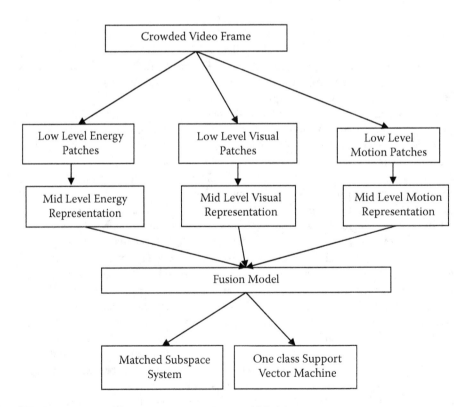

FIGURE 7.2   Overall Structure for the Proposed Model.

distinguish anomalies from crowded videos. For evaluation of the proposed algorithm, the images are taken from the UCSD data set.

7.12.1 STAGE 1

For introducing deep belief networks, one of the existing algorithms i.e., conventional restricted boltzmann machine (CRBM) [25] is used to train and learn the deep representations for images. It consists of two layers: input layer and the hidden layer. The input layer comprises binary values of detectable units. The hidden layer intern comprises two different layers, i.e., detecting layer and then a pooling layer. Each layer has N groups of units. Every group of detection layer consists of a $K_h \times K_h$ array unit to each group of pooling layer includes $K_p \times K_p$ units of p. The detection layer was divided into B*B blocks. Those are connected to one pooling layer, p_a. The energy function for CRBM is demarcated as follows:

$$E(v, h) = -\sum_{i,j} \left(h_{i,j}^k (Z^\sim \times v)_{i,j} + C_k t_{ij}^k\right) - a \sum_{i,j} v_{i,j} \qquad (7.1)$$

where C_k is shared bits of detection layer and h is the weight for flipping the matrix both vertically and horizontally.

By using CRBM, DBN is formed to learn mid-level features. The principal spatiotemporal energy is used to characterize low-level gesture features. The vitality at every pixel is $X = (x, y, t)$ and could be calculated by 3D Gaussian filters:

$$St0_\theta(x) = \sum_{x \pounds \Omega} (G_\theta^3 \times V) \qquad (7.2)$$

where G_θ^3 is the Gaussian filter and V is the input video.

7.12.1.1 Deep representation of mid-level features

The low-level features of the images are shown in Figure 7.3. Those low-level features are used to learn middle-level features of crowded videos. All of the visual features are normalized in the range of [0, 1]. Figures 7.2(a) to 7.2(f) consist of six different energies of the captured image.

7.12.1.2 Multiscale motion mapping with deep belief networks

By considering three different patches in Figure 7.4, in this, the walking man present a high speed in the red patch, normal speed in the blue patch and green patch contains a learning post. Figures 7.4(a) to 7.5(c) show energy distribution for different patches.

In this, we proposed multiscale energy for mapping different energy speeds. In this binary image is taken into consideration, three scale energy maps are used to map a different energy where each of the channels is mapped with a binary image of the same size. In this two different threshold values like T1 and T2 always choose the channel whose threshold value is T1 < T2. Apply the normalization process to map all vitalities with spatiotemporal ways and then calculate the mean threshold

(a)   (b)   (c)

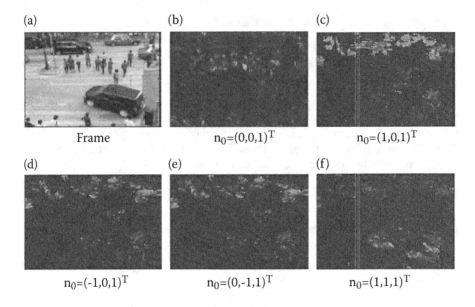

Frame $n_0=(0,0,1)^T$ $n_0=(1,0,1)^T$

(d)   (e)   (f)

$n_0=(-1,0,1)^T$ $n_0=(0,-1,1)^T$ $n_0=(1,1,1)^T$

FIGURE 7.3   Instances of Crowd Categorization Spatiotemporal-Oriented Energies Commencing the Data Set.

FIGURE 7.4   Crowded Frame.

value. If the mean threshold value is less than T1, then patch 1 foreground pixels are changed to 1, otherwise the current patch or second patch values are held based on the threshold value of the second patch.

7.12.2   STAGE 2: IDENTIFYING HIGH-LEVEL FEATURES WITH A FUSION MODEL FRAMEWORK

The current framework consists of two steps:

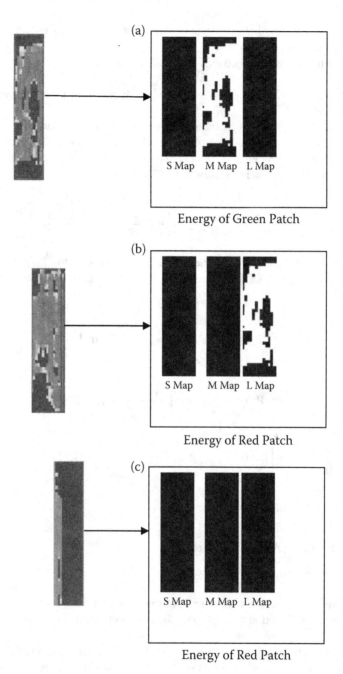

(a)

S Map   M Map   L Map

Energy of Green Patch

(b)

S Map   M Map   L Map

Energy of Red Patch

(c)

S Map   M Map   L Map

Energy of Red Patch

FIGURE 7.5 Illustration of Multi-scale Energy Maps.

1. Training of deep belief networks with motion map and obtain features from the trained crowd video
2. Learn the correlation between middle-level features

To train the network, use a supervised learning algorithm in layer by layer bit one layer by a time.

After training the first layer, the parameter of the first layer w1, b1 and b0 are stationary, and concealed value h1 of the first layer is inference. The hidden values function is input for the next layer [17]. The proposed model uses a directed graphical model with reasonable binary units. The energy for them is defined as

$$E(h_1^2, h_1^{2v}, h_1^{2map}, h_3, \theta) = \sum_{i=0}^{F} b_{1i}^2 h_{1i}^2 + \sum_{i=0}^{F} b_{1i}^{2v} h_{1i}^{2v} + \sum_{i=0}^{F} b_{1i}^{2map} h_{1i}^{2map} \qquad (7.3)$$

wherever θ is the exemplary parameter and h_3 is a bias of the hidden layer. The following formula gives the distribution of the three different sets of visible and hidden layers units.

$$P(h_{1k}^2 = 0|h^3) = \sigma\left(b_{1k} + \sum_{i=0}^{F} W_{ki} h_{1i}^3\right) \qquad (7.4)$$

$$P(h_{1k}^{2v} = 0|h^3) = \sigma\left(b_{1k}^{2v} + \sum_{i=0}^{F} W_{ki}^{3v} h_{1i}^3\right) \qquad (7.5)$$

$$P\left(h_{1k}^{2map} = 0|h^3\right) = \sigma\left(b_{1k}^{2map} + \sum_{i=0}^{F} W_{ki}^{3map} h_{1i}^3\right) \qquad (7.6)$$

where σ is the sigmoid function.

7.12.3 STAGE 3: ANOMALY DETECTION

In this, we have two different concepts:

1. Anomalies detection with match subspace system
2. Anomalies detection through one class support vector machines

Anomalies detection with match subspace system: A separate MSS is used for each layer because we deal with a 3D image and the anomalies also in $N_n \times N_m$ spatial size. In general, for detecting asubspace signal, use Gaussian formulas here; we use matched subspace detector for detection of signals in subspace.

Let X_1 represents layers L and $X_L(s)$ represents pixel at X_1 for each pixel we create row stacking and column vector. For this, we define two hypothesis values as follows:

Let $m_1(s)$ be a vector of size $N_n \times N_m$ and $\phi_1(s)$, and $\theta(s)$ be the weight vectors.

$$H_0 = S_1\phi_1(s) + m_1(s) \qquad (7.7)$$

$$H_1 = H_1\theta_1(s) + S_1\phi_1(s) + m_1(s) \qquad (7.8)$$

$$H0 = \phi_1(s) = P_0 \sum_{vi} {}^{-1/2} n_1(s) \qquad (7.9)$$

$$H1 = P_1 \sum_{vi} {}^{-1/2} n_1(s) \qquad (7.10)$$

where

$$P_0 = \left(S_i^T \sum_{vi} {}^{-1} S_1 \right)^{-1} S_L^T \sum_{vi} {}^{-1/2} \qquad (7.11)$$

$$P_1 = \left(S_i^T \sum_{vi} {}^{-1} S_1 \right)^{-1} S_L^T \sum_{vi} {}^{-1/2} [S_1H_1] \qquad (7.12)$$

The decision is framed by either P_0 or P_1.

7.13 ANOMALY DETECTION WITH ONE CLASS SUPPORT VECTOR MACHINES

One class support vector machine is used to detect anomalies in a crowded video and the one SVM delinquent could be identified using the subsequent formula:

$$Min_{a,b,c}\|a\|/2 - C + 1/nN \sum_{i=1}^{N} bi \qquad (7.13)$$

where a is learning weight vector and c is offset. N is the magnitude of the data set and b is the slack variable for patch i. The comparison of anomaly detection of the suggested method with MST is exposed in Figure 7.6.

7.14 RESULTS

In this evaluation, we consider three different levels: patch level, frame level and pixel level.

In frame level, each and every frame is taken into consideration for detecting anomalies. If at least one pixel in the frame is an anomaly, it will identify as an anomaly frame.

(a)   (b)   (c)

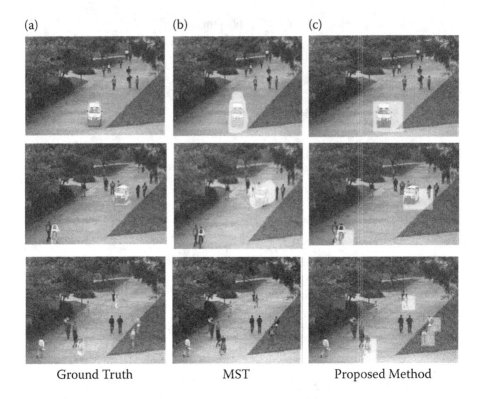

Ground Truth   MST   Proposed Method

FIGURE 7.6   Examples of Abnormal Detection on the Data Set.

Pixel Level: In this level, the fallouts are associated with the pixel of every frame. If more than 50% of the anomalies pixels were recognized, then the setting has to be considered anomalies.

Patch Level: In the patch level, we focus on true and false measurements. When more anomalies are identified, that one is treated as a true positive, or else treated as a false positive. The receiver operating characteristic curve (ROC) is cast off to the extent the accurateness of the detected anomalies. The ROC curve comprises MPR and SPR:

$$MPR = \frac{True\ Positive}{True\ Positive + False\ Negative} \qquad (7.14)$$

$$SPR = \frac{False\ Positive}{True\ Negative + False\ Positive} \qquad (7.15)$$

The performance is measured with area below curve, equal error rate (EER) and equal detected rate using the following formulas (Figure 7.7, Table 7.1):

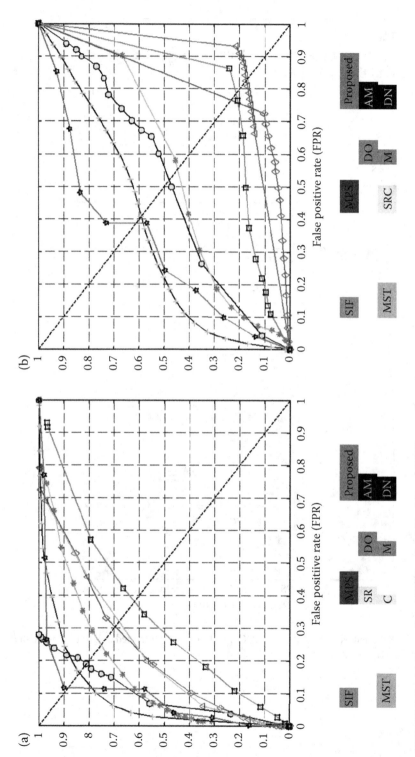

FIGURE 7.7 (a) Frame-Level ROC and (b) Pixel-Level ROC.

TABLE 7.1

Assessment EER and AUC Values of Suggested Algorithm with the Existing Algorithms

Algorithm	EER	AUC
SIF	41%	68.96%
MPS	39%	76.85%
MST	26%	83.45%
SRCS	19%	86.78%
Proposed	12.34%	91.6%

$$EER = 1 - MPR \text{ for Frame Level} \qquad (7.16)$$

$$EDR = 1 - EER \text{ for Pixel level} \qquad (7.17)$$

See Figures 7.8 and 7.9 and Table 7.2.

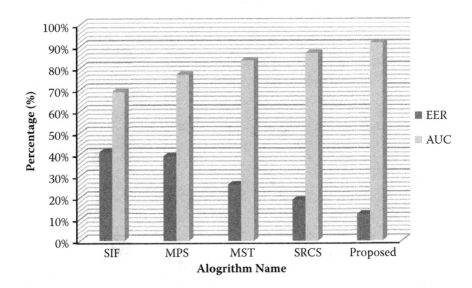

FIGURE 7.8 Analysis of EER and AUC Ideals of the Suggested Algorithm with the Existing Algorithms.

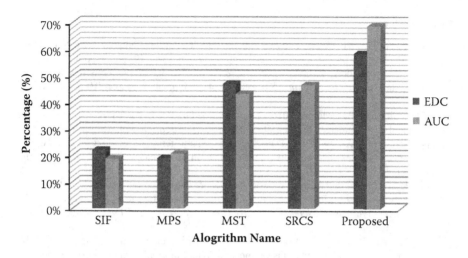

FIGURE 7.9 Analysis of EDC and AUC values of the Suggested Algorithm with the Existing Algorithms.

TABLE 7.2

Comparison EDC and AUC Values of the Suggested Algorithm with the Existing Algorithms

Algorithm	EDC	AUC
SIF	22%	18.7%
MPS	19%	20.67%
MST	47%	43.23%
SRCS	43%	46.56%
Proposed	58.49%	68.74%

7.15 CONCLUSIONS & FUTURE SCOPE

In this study, present the novel approach for anomaly detection of crowd video using deep belief networks (DBNs). In this, low-level features are extracted from spatiotemporal measurement, middle-level features are extracted using three conventional mapping machines and high-level features are extracted from the fusion model. Match subspace system and one class support vector machines are used to distinguish anomalies from crowded videos and 3D images. The proposed algorithm is suitable for identifying and detecting anomalies in images and videos. It was implemented with an available data set and measures the EER and EDR values, and it produces better results with various existing algorithms. In the future, more concentration has to be made to construct more than deep neural network architecture and different multi-modal, fusion techniques for anomaly identification and detection for more complex videos.

REFERENCES

[1] X Song, M Wu, C Jermaine & S Ranka 2007. 'Conditional anomaly detection', IEEE *Transactions on Knowledge and Data Engineering*, vol. 19, no. 5, pp. 631–645.

[2] Varun Chandola, Arindam Banerjee & Vipin Kumar 2009. 'Anomaly detection: A survey', *ACM Computing Surveys*, vol. 41, no. 3. https://doi.org/10.1145/1541880. 1541882

[3] K Sitara & BM Mehtre 2016. 'Digital video tampering detection: An overview of passive techniques', *Science Direct-Digital Investigation*, vol. 18, pp. 8–22.

[4] Nannan Li, Xinyu Wu & Dan Xu 2015. 'Spatio-temporal context analysis within video volumes for anomalous-event detection and localization', *Neurocomputing*, vol. 155, pp. 309–319.

[5] Pedro Canotilho Ribeiro, Romaric Audigier & Quoc Cuong Pham 2016. 'RIMOC, a feature to discriminate unstructured motions: Application to violence detection for video-surveillance', *Science Direct - Computer Vision and Image Understanding*, vol. 144, pp. 121–143.

[6] Dan Xu, Yan Yan & Elisa Ricci 2016. 'Detecting anomalous events in videos by learning deep representations of appearance and motion', *Science Direct Computer Vision and Image Understanding*, vol. 156, pp. 117–127. https://doi.org/10.1016/j.cviu.2016.10.010

[7] Andrea Pennisi, Domenico D. Bloisi & Luca Iocchi 2016. 'Online real-time crowd behavior detection in video sequences', *Computer Vision and Image Understanding*, vol. 117, no. 10, pp. 1436–1452.

[8] Ying Zhang, Huchuan Lu & Lihe Zhang 2016. 'Video anomaly detection based on locality sensitive hashing filters', *Science Direct Pattern Recognition*, vol. 59, pp. 302–311.

[9] Liyang Yu, Huanran Wang & Qi Han 2016. 'Exposing frame deletion by detecting abrupt changes in video streams', *Science DirectNeurocomputing*, vol. 205, pp. 84–91.

[10] V. Mahadevan, Weixin Li, Viral Bhalodia & Nuno Vasconcelos 2010. 'Anomaly detection in crowded scenes', *Proceedings of the IEEE Conference on CVPR*. https://doi.org/10.1109/CVPR.2010.5539872

[11] A. A. Sodemann, M. P. Ross & B. J. Borghetti 2012. 'A review of anomaly detection in automated surveillance', *IEEE Transactions on Systems, Man, and Cybernetics, Part*, vol. 42, no. 6, pp. 1257–1272. https://doi.org/10.1109/TSMCC.2012.2215319

[12] M. Sabokrou 2015. 'Realtimeanomaly detection and localization in crowded scenes', *Proceedings of the IEEE Conference on CVPRW '1*, USA, June.

[13] A. Del Giorno 2016. 'A discriminative framework for anomaly detection in large videos', *Proceedings of the ECCV'16*, pp. 334–349, Springer, Cham.

[14] Irene Cramer & Prakash Govindarajan 2018. 'Detecting anomalies in device event data, Bosch Software Innovations GmbHoT Analytics', *IoT Analytics*, Ullsteinstraße 128, 12109 Berlin, Germany.

[15] Salima Omar, Asri Ngadi & Hamid H. Jebur 2016. 'Machine learning techniques for anomaly detection', *International Journal of Computer Applications*, vol. 79, no. 2. https://doi.org/10.5120/13715-1478

[16] Markus Goldstein & Andreas Dengel 2012. Histogram-based outlier score (hbos): A fast unsupervised anomaly detection algorithm, KI-2012: Poster and Demo Track, pp. 59–63.

[17] Sukadev Meher 2012. 'A trajectory-based ball detection and tracking system with applications to shot-type identification in volleyball videos', *International Conference on Signal Processing and Communications*, Bangalore, India. https://doi.org/10.1109/SPCOM.2012.6290005

[18] Nannan Li 2015. 'Anomaly detection in video surveillance via gaussian process', *International Journal of Pattern Recognition and Artificial Intelligence*, vol. 29, no. 6, p. 150426191333005. https://doi.org/10.1142/S0218001415550113

[19] Wenjie Hu, Yihua Liao & V. Rao Vemuri 2003, June. 'Robust anomaly detection using support vector machines'. In Proceedings of the international conference on machine learning. pp. 282–289.

[20] Sheeraz Ari, Wang Jing, Hussain Fida & Fei Zesong 2019. Trajectory-Based 3D Convolutional Descriptors for Human Action Recognition. Journal of Information Science & Engineering, vol. 35, no. 4. https://doi.org/10.13140/RG.2.2.25039.12969

[21] Yingchi Mao 2017. 'An adaptive trajectory clustering method based on grid and density in mobile pattern analysis', *Sensors*, vol. 17. https://doi.org/10.3390/s17092013.

[22] Yadigar Imamverdiyev 2017. 'An anomaly detection based on optimization', *International Journal of Intelligent Systems and Applications*, vol. 9, no. 12, pp. 87–96.

[23] Keith Hollingsworth 2018. 'Energy anomaly detection with forecasting and deep learning', *IEEE International Conference on Big Data (Big Data)*. https://doi.org/10.1109/BigData.2018.8621948

[24] Luis Martí, Nayat Sanchez-Pi, José Manuel Molina & Ana Cristina Bicharra Garcia 2014. 'Yet Another Time Series Segmentation Algorithm for Anomaly Detection in Big Data Problems', *International Conference on Hybrid Artificial Intelligence Systems HAIS*. https://doi.org/10.1007/978-3-319-07617-1_61

[25] Shen Su 2019. 'A correlation-change based feature selection method for IoT equipment anomaly detection', *Applied Sciences*, vol. 9, p. 437. https://doi.org/10.3390/app9030437

[26] Chunyu Chen, Yu Shao & Xiaojun Bi 2015. 'Detection of anomalous crowd behavior based on the acceleration feature', *IEEE Sensors Journal*, vol. 15, no. 12. https://doi.org/10.1109/JSEN.2015.2472960

[27] Hemank Lamba & Thomas J. Glazier 2017. 'A model-based approach to anomaly detection in software Architectures', *ACM*. https://doi.org/10.1145/1235

[28] Kelathodi Kumaran Santhosh, Debi Prosad Dogra & P.P. Roy 2020. 'Anomaly detection in road traffic using visual surveillance: A survey', ACM Computing Surveys (CSUR), vol. 53, no. 6, pp. 1–26.

[29] Olga Isupova, Danil Kuzin & Lyudmila Mihaylova 2016. 'Dynamic hierarchical Dirichlet process for abnormal behaviour detection in video', 19th *International Conference on Information Fusion (FUSION)*, Heidelberg, Germany.

[30] Huihui Zhu, Bin Liu, Yan Lu, Weihai Li & Nenghai Yu 2018, December. Real-time Anomaly Detection with HMOF Feature. In Proceedings of the 2018 the 2nd International Conference on Video and Image Processing. pp. 49–54.

[31] Vinod Nair & James Clark 2002. 'Automated visual surveillance using hidden Markov models', *International Conference on Vision Interface*. www.cse.psu.edu/~rtc12/CSE586Spring2010/papers

[32] Amogh Mahapatra 2017. 'Contextual anomaly detection in text', *Data Algorithms*, vol. 5, no. 4, pp. 469–489.

[33] Amit Adam 2008. 'Robust real-time unusual event detection using multiple fixed-location monitors', *IEEE Transactions on Pattern Analysis and Machine Intelligence*, vol. 30, no. 3. https://doi.org/10.1109/TPAMI.2007.70825

[34] Xuguang Zhang 2018. 'Energy level-based abnormal crowd behavior detection', *Sensors (Basel)*, vol. 18, no. 2, pp. 423. https://doi.org/10.3390/s18020423

[35] R. Mehran, A. Oyama & M. Shah 2009. 'Abnormal crowd behavior detection using social force model', *Proceedings of the IEEE Conference on Computer Vision and Pattern Recognition (CVPR '09)*, pp. 935–942, Miami, Fla, USA.

[36] Venkatesh Saligrama, Janusz Konrad & Pierre-Marc Jodoin 2010. 'Video anomaly identification', *IEEE Signal Processing Magazine*, vol. 27, no. 5. https://doi.org/10.1109/MSP.2010.937393

[37] Yogita Dhanraj Mistry & Dilip Ingole 2014. 'Content-based image retrieval using DWT based feature extraction and texture, shape and color features', *International Journal of Research in Computer and Communication Technology*, vol. 3, no. 11.

[38] Claudio Piciarelli, Christian Micheloni & Gian Luca Foresti 2008. 'Trajectory-based anomalous event detection', *IEEE Transactions on Circuits and Systems for Video Technology*, vol. 18, no. 11. https://doi.org/10.1109/TCSVT.2008.2005599

[39] Chao Dong, Chen Change Loy, Kaiming He & Xiaoou Tang 2016. 'Image super-resolution using deep convolution networks', *IEEE Transactions on Pattern Analysis and Machine Intelligence*, vol. 38, no. 2. https://doi.org/10.1109/TPAMI.2015.2439281

[40] Yong Shean Chong & Yong Haur Tay 2015. 'Modelling video-based anomaly detection using deep architectures: challenges and possibilities', *IEEE Conference on Asian Control Conference*, pp. 1–8.

[41] Jiahui Wen, Mingyang Zhong & Zhiying Wang 2015. 'Activity recognition with weighted frequent patterns mining in smart environments', *Journal of Expert Systems with Applications*, vol. 42, no. 17–18, pp. 6423–6432.

8 Innovation in Multimedia Using IoT Systems

Abdullah Ayub Khan[1,2], Asif Ali Laghari[1], Aftab Ahmed Shaikh[1], Zaffar Ahmed Shaikh[2] and Awais Khan Jumani[3]

[1]Faculty of Computing Science, Sindh Madressatul Islam University, Karachi 74000, Sindh, Pakistan
[2]Faculty of Computing Science and Information Technology, Benazir Bhutto Shaheed University Lyari, Karachi 75660, Sindh, Pakistan
[3]Department of Computer Science ILMA University, Karachi, Pakistan

8.1 INTRODUCTION

Nowadays, the widespread information technology claim that the new concepts, frameworks, models, methods, architectures, tools and mechanisms of communication technology are driving processes of multimedia systems and design changes and leading to the creation of the Internet of things (IoT)–enabled innovative multimedia systems [1]. The approach adaptation and design are used only when understanding the social and technological environment and their needs, whereas shaping and examining the approach processes and manipulates the degree of level accordingly. This study addresses the IoT-based latest innovation in multimedia systems and design; the context of the Internet of multimedia Things and cloud service deliverance; and the processes of social, cultural, political and economic changes throughout the world in the year 2019–2021 [2].

In the past few decades, information and communication technology are gaining rapid advancement in multimedia systems and designs that are more robust and perform far better than the previous one; moreover, they innovate some areas such as smart homes, event-based event processing, cloud-enabling technologies and 0IoT [3]. The Internet of multimedia Things (IoMT) involves the fast-growing cloud-based service delivery network with interconnected things that collect, examine, analyze and exchange data using a variety of embedded sensors such as remote management of mobile-based sensors [4]. The sensor is a subsystem attached with the Internet of multimedia systems to detect and check changes in the environment, including light, proximity, temperature and other environmental senses incorporated [5].

DOI: 10.1201/9781003196686-8

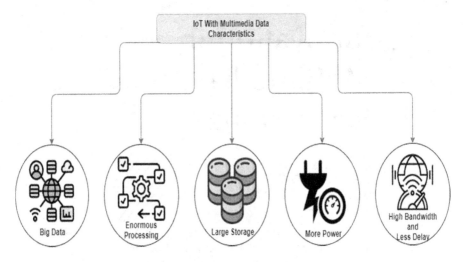

FIGURE 8.1   IoT with Multimedia Data Characteristics.

And so, each sensor is embedded with some critical computational intelligence in which the handling and monitoring system behaviour and measure the functionality. However, cloud pay-per-use platforms provide scalability, elasticity, easy access, storage space, computing power, application development environment and maintainability of IoMT. The composition and control of the advanced IoMT are only possible by rule-based event processing [6] (Figure 8.1).

In the business intelligence context, mobile marketing, live broadcasting, live casting, podcasting, audio, image, video, data, and filesharing poses critical challenges throughout the communication between small businesses, customers and employees. The current multimedia system is able to communicate mass marketing strategies like the choice on demand to the devices [7]. The statistical forecasting tools conjecting with multimedia tools to support business and their revenue. And so, the marketing especially, the sales and promotions of their products and services. However, the Internet of multimedia Things (IoMT) robust productivity and the marketing promotion strategies, such as offer real-time incentives, join-like minded community, automatic drive traffic to your website and webcasting [8]. Through these, emerge some critical issues in the current system of IoMT when creating multimedia contents that are listed as follows: (Figure 8.2).

- broadcast (show how to use your product)
- extensibility (expend according to client request)
- entertain (customer satisfaction)
- live cast (provide a unique service)

8.1.1   OBJECTIVE OF THE STUDY

This chapter addresses the hot topic of IoT in multimedia systems and design, a new way of operational control and monitor smart devices using cloud-enabling

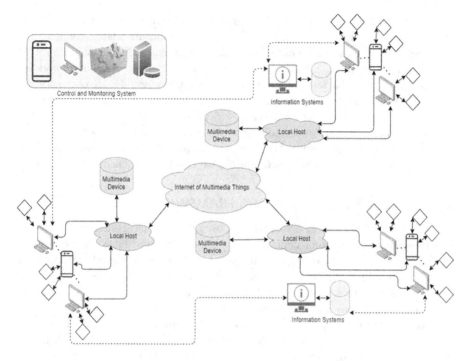

FIGURE 8.2 The Current Running Internet of Multimedia Things Framework.

technology. The main objectives and contribution of this chapter are discussed as follows:

- Studied various current systems, the latest innovation in multimedia, and impact of IoT-based solutions and the real-time applications of IoMT are explained.
- Multimedia operation and control monitoring IoT-enabled solution detect system behavior while communicating from one to another end in the channel. And so, analysis the packet and node segmentation during the transmission such as audio, photo, video, live casting and filesharing.
- Investigate the significance of IoT-based multimedia systems and design frameworks and highlight the importance of network sensors and compression.
- In the real-time environment, classify the vital role of IoMT in smart home, health, agriculture and commercial cloud gaming streaming, and others.
- Also proposed a novel and secure service delivery framework for the Internet of multimedia Things and presented the working mechanism of the proposed framework through the sequence diagram along with critical implementation challenges.

The rest of this chapter is organized as follows. The next section discussed the IoT-based multimedia frameworks and their radical impact on information and communication technology. Section 3 focuses on the traditional and current trends in IoT-enabled multimedia applications, especially the Internet of multimedia Things for real-time streaming, live broadcasting, commercial cloud-based gaming and multimedia things for smart health and agriculture. In Section 4, the proposed secure and novel framework for IoT-enabled multimedia service delivery is presented. Further, this demonstrates the working mechanism of the proposed framework through a sequence diagram and highlights critical implementation challenges. Section 5 describes IoT security-related limitations and issues in multimedia systems, including network behavior, video streaming and image processing. Open research areas and the future direction of the technology are discussed in Section 6. Finally, we conclude our chapter in Section 7.

8.2 IOT-ENABLED MULTIMEDIA FRAMEWORKS

Billions of physical systems are connected over the Internet, collect and share in-structions, small computing chips use a wireless network and a unique way to handle devices ubiquitously [9]. Distinct objects are connected and add sensors with a level of digital intelligence, a practice to make devices intelligent because of this, enable communication with different devices and share data in the real-time en-vironment without any need for intervention. Collaborate with technology in order to produce better products such as IoT and sensors-based multimedia systems, IoT-enabled cloud and fog computing and security. Various related literature appears, merge two or more technologies and provide new ways of information and com-munication technology to robust performance in terms of system functionality, privacy and efficiency [10] (Table 8.1).

TABLE 8.1
Related Work of IoT-Enabled Multimedia Systems Frameworks

Research Title	Proposed Framework	Features	Reference
Towards 6G Architecture for Energy Efficient Communication in IoT-Enabled Smart Automation Systems	A. H. Sodhro et al. focuses on energy-efficient communication and quality of experience (QoE) framework for capturing user experience through user terminal devices during multimedia-based communication transmission.	• Improve system to check level of user satisfaction • Entropy optimization • 6G-driven multimedia data • Used correlation model	[11]
An efficient framework using visual recognition for IoT-based smart city surveillance	M. Kumar et al. proposed efficient subspace fast decomposition transformation for IoT-based smart city surveillance system using Chi-Square.	• Visual recognition using local binary pattern • High dimensionality reduction • Battery optimistic	[12]

TABLE 8.1 (Continued)

Related Work of IoT-Enabled Multimedia Systems Frameworks

Research Title	Proposed Framework	Features	Reference
An IoT Framework for Heart Disease Prediction Based on MDCNN Classifier	M. A. Khan proposed an IoT-based multimedia framework that evaluates heart disease using a modified deep convolutional neural network. Through the monitoring, a smartwatch and heart monitor devices are attached with the patient that analyzes blood pressure and pluses in real time.	• Robust prediction systems • Accurate heart disease diagnosing • Efficient classifier	[13]
An Efficient Named Data Networking-Based IoT Cloud Framework	X. Wang and S. Cai present an efficient IoT-based cloud data networking framework. In this, Internet protocol-based delivery method to retrieve data using mobile devices.	• Evaluate quantitatively • Limited flooding • Unicast to acquire data • High success rate of data retrieval and reduce cost	[14]
A Trust Computed Framework for IoT Devices and Fog Computing Environment	G. Rathee et al. present a secure routed and handoff framework used to avoid cyberattacks and exploring trust between fog nodes in the cloud environment using IoT-enabled communication of multimedia behaviour.	• Enabled trust between fog and IoT layer • Detect malicious • Conventional security • Mobility	[15]
A Lightweight Replay Detection Framework for Voice Controlled IoT Devices	K. M. Malik et al. proposed a framework for IoT-based voice-controlled, lightweight distortion detection using acoustic ternary patterns-gammatone cepstral coefficient features and SVM for classifying distinct patterns.	• Multimedia enabled voice-activated services • Detect high level harmonic distortion • Replay malicious attacks	[16]

8.2.1 Compression in Wireless Network Sensing

In a digital communication network, a wireless sensor node consumes more power to communicate with different units. Several methods are proposed, whereas a data compression mechanism is used to reduce the size of embedded data then exchange it on the network. As a result, a large amount of energy saves [17]. R. Sheeja et al. elaborate the concept of network sensor compression, the higher ratio of information may reduce the percentage of power consumption [18]. In this regard, S. Kalaivani et al. discussed some simple data compression algorithms for multimedia-based wireless sensor networks [19]. A simple lossless data compression used multi-channel Huffman coding to reduce wireless sensor networks data and save power. However, signal compression is one of the significant aspects of communication and reduce the cost of communication and robust deployment life cycle for multimedia-enabled information sharing and smooth communication with the use of the wireless sensor networks.

8.3   IOT-BASED MULTIMEDIA APPLICATIONS

The Internet of multimedia Things is associated with the Internet protocols, distinct interfaces and representation of multimedia-related information communication [20]. The multimedia-enabled service delivery and applications are built for human interactions such as point-to-point and device-to-device communication in the virtual and physical environment. In this chapter, we have focused on IoMT-based real-time streaming, i.e., mostly on gaming, smart health and agriculture-related things, the detailed discussion of the mentioned applications are as follows:

8.3.1   INTERNET OF MULTIMEDIA THINGS FOR REAL-TIME STREAMING

In real-time streaming, multimedia service delivery and application development activities rely on several preliminary requirements such as bandwidth, storage, high data processing units, latency, etc. [21]. Moreover, the IoT weaknesses in terms of dedicated protocols and security issues in a multimedia environment. IoMT emerges as a more efficient information and communication technology, a new paradigm that plays a significant role in cloud gaming, for example, live streaming through YouTube and cloud-game platforms are Nware, Vortex, Stadia and NetBoom [22]. Several commercial and non-commercial clouds enabled live gaming platforms that facilitate users to pay-as-you-play services [23]. Nevertheless, the Internet of multimedia-based cloud gaming obtained results that proved the outperformance and the smoothness in real-time streaming. IoMT achieved the basic once-in-the-live streaming such as end-to-end delay, the ratio of package delivery, data compression and save power consumption, throughput and delivery services acknowledgment. In short, IoMT provides better performance than the other state-of-the-art protocols.

8.3.2   INTERNET OF MULTIMEDIA THINGS FOR SMART HEALTHCARE

The IoMT integrates distinct concepts of computing technology that enabled systems, cloud computing and intelligent sensors in a single domain. Also, it deals with all the scalar properties and multimedia communications [24]. New technologies, methods, approaches and architectures proposed for fast peace of human life, more advanced and sophisticated smart healthcare frameworks proposed individual according to the health requirement highlighted by S. U. Amin and M. S. Hossain in 2020 [25]. These proposed frameworks based on IoMT devices are highly digitalized, embedded smart and more powerful sensors along with 5G network and used ubiquitous fog and edge computing. An intelligent solution reduces the size of data and compressed signal transmission and saves power consumption for real-time health-related data monitoring and control based on classified data. The system speeds up health diagnosis to live tracking, monitoring and control and provides the more efficient and accurate treatment [24,25]. The real-time state-of-the-art IoMT for smart healthcare platform are as follows:

- Virtual hospitals
- Wearable biosensors
- Smart thermometer
- Connected inhalers
- Smart watch monitoring
- Automated insulin delivery system
- Assisting the elderly
- The privacy conundrum

8.3.3   INTERNET OF MULTIMEDIA THINGS FOR AGRICULTURE

In this era, the research work on smart agriculture and robust production is an area of concern for all researchers of computing technology, when considering managing, monitoring, weather conditions, soil quality and production of crops [26,27]. The IoMT in agriculture changes research dynamics drastically with real-time analysis of agricultural-based sensors that are placed on the crops and collect data. As a result, the smart monitoring of crop growth and fast production reduce energy, waste of resources, time, farmer efforts and low production risk [28]. Moreover, the preservation of collected data and analyzed results are stored in the cloud environment (such as pay-as-you-go services) and automate the process by integrating sensors data using the Internet of Things and applied actions. Several IoMT-enabled smart monitoring and intelligent agricultural systems include greenhouse automation, crop management and monitoring of climate conditions, IoT-based cattle monitoring system, precision farming, agricultural drones, predictive analytics and end-to-end farming management systems [29,30].

8.3.4   INTERNET OF MULTIMEDIA THINGS FOR SMART HOME AND ENERGY MANAGEMENT

A residential extension involves embedded intelligent chips and automated technology for monitoring and controlling a smart home and its energy system. The residential extensions are smart televisions, air conditions, lighting, kitchen-based appliances, washers, refrigerators and other security and surveillance systems. The communication between these devices is controlled remotely through mobile phones, the Internet and a time schedule. Although these systems are connected with switches and a centralized hub, there are also embedded wireless sensors to control home devices and wall-mounted ubiquitous devices connected to cloud services. The Internet of the multimedia smart home provides energy efficiency and reduces costs and increases security, interoperability and convenience. IoMT-enabled systems are adaptive and adjustable because of a flexible infrastructure for whenever an ongoing change is required. Moreover, the architecture enables microcontrolled wireless sensors, multimedia devices to monitor overall activity, instrumental data processes and actuators for actions.

8.3.5 INTERNET OF MULTIMEDIA THINGS FOR A SMART CITY

In IoMT, the smart city is a framework composed of a collaboration of information and communication technology, the ability to remote monitor, track and trace and control related devices. In doing so, this will initiate the latest insights and direct attempt action on the information from huge streams of real-time data. The integration of a high degree of application resources of information systems is an essential technology for the development of urban areas such as smart industry, services, management, technology and life. The main features of smart cities are to be instrumented, interconnected and intelligent. The installation of IoT-based sensors with everything may connect to the cloud services through specific protocols. In this regard, the system provides a secure and integrated environment to exchange information and communication between nodes in order to achieve monitoring, location, tracking and land recognition. The massive growth of information technology, especially in the development of the smart city with IoMT applications, creates many scientific limitations from both the industry and academic side.

8.4 PROPOSED FRAMEWORK FOR IOT MULTIMEDIA SERVICE DELIVERY

Figure 8.3 presents the proposed cloud-enabled Internet of a multimedia service delivery framework. The deployment of this solution requires a two-step setup, one from integrated sensors and multimedia devices and user communication to the Internet. And so, the second from the Internet connection to the cloud-enabled service delivery systems, whereas the Internet performs bridge connectivity between them, either it is connected through guided (wired) or unguided media (4G/5G). To access IoMT service delivery, publication subscription and valid authentication are required. After the authentication process, the cloud service provider provides services to the users with IoMT-based software, platform and infrastructure. The cloud manager is responsible for overall cloud monitoring, control, resources management, risk management, storage and multimedia service directory management.

Through the secure preservation of data over the cloud, we have maintained distributed storage for making novel and secure IoMT service delivery node transactions, and also provided privacy and security in terms of information integrity, confidentiality and secure control access of IoMT systems, as shown in Figure 8.4. However, the multimedia service directory manages several applications

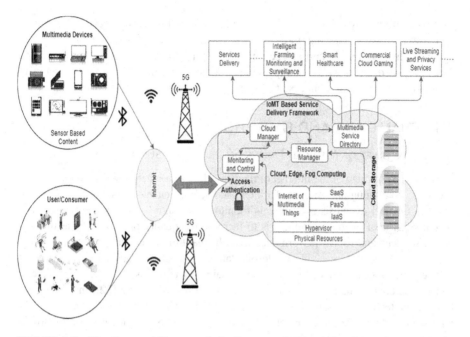

FIGURE 8.3 The Proposed Framework for Internet of Things (IoT) Enabled Multimedia Service Delivery.

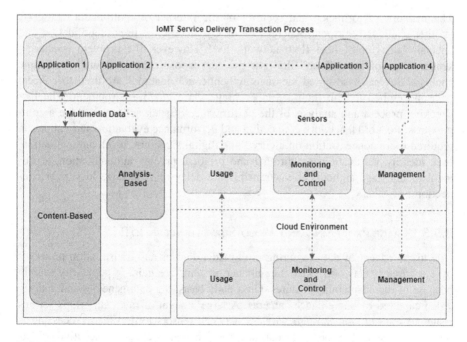

FIGURE 8.4 IoMT Service Delivery Node Transaction Execution Process.

in a concurrent way. The services are handled and delivered by service directories that are intelligent farming monitoring and surveillance, smart healthcare management, live streaming and privacy management, cloud gaming and many others. This proposed framework also provides robust performance and more security compared to the other related state-of-the-art architectures.

8.5 CHALLENGES AND LIMITATIONS

8.5.1 INTERNET OF MEDICAL THINGS SECURITY AND SUSTAINABILITY

With the rapid growth of the Internet of medical Things and their association with cloud, fog, edge computing and multimedia integrated sensors enabled systems, the direction turns to medical diagnosis and patients monitor and control systems themselves. The fundamental goal of medical things is to increase the quality of peace of life through health services and robust performance [31]. In real-time healthcare management, the Internet of medical Things collects sensory data from a patient's analyses and processes it. The wide range of data collection and maintaining integrity is a challenging problem, and so, the confidentiality of health data transmission and security [32]. In this regard, blockchain technology can solve such types of interoperability in the medical environment and provide multi-model secure data transmission of patients' information in the healthcare sector [33].

8.5.2 IoMT NETWORK CONNECTION AND PERFORMANCE MATRIX

The network performance and security of the Internet of multimedia Things is another challenging aspect, whereas privacy protection of user identification and analysis behavior of host IPv6 network [34]. However, the problem is sharing sensory data of the IoT and the user privacy exposure information in real-time analysis. Several hash-based message authentication methods are used to conceal data and embed it with multimedia coverage via a cloud environment. The list-checking process and analysis of the performance of such activity is quite a complex task. For IoMT network connection and performance evaluation and validation required more concentration on a curve-like digital signature, batch authentication, bulk identity verification, conditional and protocol privacy authentication, a protocol for message authentication, transmission delay rate, package loss, peer connection and others.

8.5.3 IMAGE PROCESSING AND VIDEO SURVEILLANCE IN IoT

Due to the advent of data-capturing multimedia devices and transmission protocols over a wireless network, video surveillance systems are gaining popularity rapidly because of the adaptability features. On a daily basis, big data generate and analyze video captured by surveillance cameras. A large amount of raw data brings serious issues to transmission through wireless network sensors [35]. Video processing is required to automate procedures that analyze redundant frames, consecutive objects of interest and filtration before uploading and transmission over the Internet [36].

There are two main constraints that need to be considered for a feasible solution: one is a frame filtering module by dynamic background and collaborative validation [35]. Furthermore, the study needs to understand this from a different perspective and find a relative research direction.

8.5.4 Outdated Software Versions and Hardware Devices

With the increased use of the Internet of multimedia Things, IoMT manufacturers and service providers focus on the development of digital devices and building a new series, whereas security is one of the open areas that needs concentration, paying a significant amount of attention to create a novel and secure architecture for both the software and hardware. The majority of multimedia-enabled devices do not get enough updates. Initially, the systems are secure, but over time they require updates, patches to the system's security and security from vulnerable attacks or harm to devices. The manufacturing companies are responsible for release regular updates, and customers have to update devices whenever the patch is released; if it is not updated, this may lead to data breaches.

8.5.5 Autonomous Systems for Data Management (Monitoring and Control)

From the network perspective, a massive amount of data is generated from different connected multimedia devices. Handling this large bulk of data is a challenging problem nowadays. Several artificial intelligence (AI) tools and methods are proposed to investigate and analyze; there is no such set of rules defined for analyzing big data and pattern detection of multimedia data from IoT and network experts. Nevertheless, the use of AI tools will be risky; for example, the configuration leads to complexity, requiring more computational power, consuming more energy and power, lacking integrity and confidentiality and third-party authentication.

8.6 OPEN RESEARCH ISSUES

8.6.1 IoMT Privacy and Security

The IoMT is an ecosystem of connected sensors, wearable multimedia devices, monitoring and control systems [37,38]. This technology makes several applications, such as live cloud-based gaming stream, smart health, intelligent farming and others that reduce systems' cost, time, efforts and provide efficient responses and increase productivity [39,40]. With the advent of cloud computing, wireless and smart sensors, ubiquitous devices, pervasive computing and big data analysis, IoMT is transforming the systems by delivering services and enabling seamless communication of multimedia data [41]. At this point, IoMT provides various benefits, but the systems require serious concern on security and privacy [42]. The IoMT applications collect and process critical information and, according to this, it makes decisions based on the information. Moreover, secure such sensitive information and personal details, systems required robust security technology that restricts unauthorized access and reduces potential risks and vulnerabilities [43]. And so, combining multiple technologies such as blockchain hyperledger-enabled IoMT will provide a more efficient solution in terms of privacy and prevent integrity and confidentiality of multimedia data.

8.6.2 Lack of Standardization and Platform Interoperability

The cloud service provider (CSP) plays an important role to manage the standard of cloud services and service-level policies, including user authenticity, resource elasticity, scalability, platform maintainability, safety security, data secrecy and integrity, transparency, portability, sharing, exchange packages and effectiveness between CSP and cloud users [44]. The CSP looks after the strategy to manage extraction, preservation, transfer, save and exchange IoMT-related data along with privacy and security. In this way, IoMT still required a more efficient mechanism to organize and deliver services over the cloud and a robust system security strategy by using blockchain technology [45,46]. Another challenging factor is to platform interoperability and authors found a lack of multiuser intercommunication, delivery of services, interaction and integration of distinct data collection, which conducts the meaningful transaction [42]. For instance, it is difficult to have adaptability and IoMT-based service delivery implementation between users.

8.6.3 IoMT and Quality of Experience (QoE)

The Internet of multimedia Things is a worldwide network of interconnected multimedia devices, connected through specific addresses and protocols based on wireless sensor networks [43]. This system is incorporated with various different technologies such as wireless sensors, multimedia devices, wearable monitoring and control systems and other devices so that it can communicate and, based on this interconnectivity, the systems can make a decision and perform actions [49]. Quality of experience (QoE) for IoT application evaluations, with particular involvement of multimedia contents, is quite a complex task [50,51]. First, analyze the IoT infrastructure and the influences factors that make systems' experience unreliable [52]. Then, estimate the influence factors and complete scenario of multimedia application-related contributions [53]. And most importantly, there is serious concern on the building of remote monitoring and control (QoE with the cyber-physical system) of smart healthcare, intelligence surveillance, live streaming and agricultural technology [54,55].

8.6.4 IoMT and Device-to-Device Communication

The development of information and communication technology has turned a series of changes in people's life [56]. Almost everything is becoming real and hands-on daily practices because of the development of the Internet and communication [57]. The Internet of multimedia Things is used in various applications, such as automation, intelligent farming, driverless cars and smart homes [58]. These sensory-based devices are self-configuring and capable of making decisions according to the data received from the environment. When IoT devices communicate with each other, there are several challenges that emerge, one being the trust issues between devices [59]. Several artificial intelligent machine learning and deep learning–based approaches are applied to maintain trustworthy communication between nodes and calculate the trust score [60]. But still, more gap is available in this domain, for example, a large number of distinct computing devices send and receive millions and billions of compressed data, automate analysis of the nature and behavior of a host network and their range and permission to the devices on whether to communicate or not.

8.7 CONCLUSION

This chapter introduced IoT-enabled multimedia things, such as privacy and security issues, multimedia service delivery frameworks, applications, challenges and limitations and open research areas. Several IoMT-related frameworks are proposed by different scholars, whereas few of these methods are mentioned and highlight their practical application running in a real-time environment. However, the resources of IoMT are limited, same as the issues in IoT, because of its inherent nature. The primary resources are available to the IoMT, such as memory and storage, power, processing units, network bandwidth and sensor integration. With the huge production of multimedia data by mobile technology, a quick analysis examines a massive amount of data collecting, accumulating, aggregating, storing and sharing information constantly. Continuous monitoring and control of multimedia-related gathered data and processing such a large amount of data rapidly may impact the quality, and somewhere compromised accuracy and confidentiality; for example, in smart healthcare management systems, commercial cloud gaming and streaming and intelligent farming. Therefore, controlling this massive amount of data and lack of standardization and system privacy are becoming the implementation challenges and limitations. Overall, the authors found the proposed IoMT service delivery framework to be flexible and efficient enough to be used in multimedia resource delivery through devices. In fact, we have compared our proposed framework with other state-of-the-art related multimedia services delivery systems.

AUTHOR CONTRIBUTIONS

Abdullah Ayub Khan has written the original draft and preparation; Asif Ali Laghari has reviewed, rewrote and edited; Aftab Ahmed Shaikh has investigated; Zaffar Ahmed Shaikh has designed framework; and Awais Khan Jumani has applied software tools. All authors read and agreed to the published version of this chapter.

The authors did not have any conflict of interest.

REFERENCES

[1] Wang, Y., 2017. November. On the innovation of multimedia technology to the management model of college students. In *International Symposium on Intelligence Computation and Applications* (pp. 175–182). Springer, Singapore.

[2] Ružičić, V.S. and Micić, Ž.M., 2017. Creating a strategic national knowledge architecture: A comparative analysis of knowledge source innovation in the ICS subfields of multimedia and IT security. *Computers & Security*, 70, pp. 455–466.

[3] Wang, X. and Jia, X., 2021. Mechanism and policy of multimedia communications model and green low-carbon development mechanism in coal industry with cloud Internet of Things. *International Journal of Communication Systems*, 34(6), p. e4523.

[4] Al-Turjman, F. and Alturjman, S., 2020. 5G/IoT-enabled UAVs for multimedia delivery in industry-oriented applications. *Multimedia Tools and Applications*, 79(13), pp. 8627–8648.

[5] Domb, M., 2019. Smart home systems based on internet of things. *Internet of Things (IoT) for Automated and Smart Applications*. In *IntechOpen*. https://doi.org/10.5772/INTECHOPEN.84894

[6] Zikria, Y.B., Afzal, M.K. and Kim, S.W., 2020. Internet of multimedia things (IoMT): Opportunities, Challenges and Solutions. *Sensors*, 20(8), p. 2334.

[7] Kumar, V., Saboo, A.R., Agarwal, A. and Kumar, B., 2020. Generating competitive intelligence with limited information: A case of the multimedia industry. *Production and Operations Management*, 29(1), pp. 192–213.

[8] Choi, J., Yoon, J., Chung, J., Coh, B.Y. and Lee, J.M., 2020. Social media analytics and business intelligence research: A systematic review. *Information Processing & Management*, 57(6), p. 102279.

[9] Ande, R., Adebisi, B., Hammoudeh, M. and Saleem, J., 2020. Internet of Things: Evolution and technologies from a security perspective. *Sustainable Cities and Society*, 54, p. 101728.

[10] Sestino, A., Prete, M.I., Piper, L. and Guido, G., 2020. Internet of Things and Big Data as enablers for business digitalization strategies. *Technovation*, p. 102173. https://doi.org/10.1016/j.technovation.2020.102173

[11] Sodhro, A.H., Pirbhulal, S., Zongwei, L., Muhammad, K. and Zahid, N., 2020. Towards 6G architecture for energy efficient communication in IoT-enabled smart automation systems. *IEEE Internet of Things Journal*. https://doi.org/10.1109/JIOT.2020.3024715

[12] Kumar, M., Raju, K.S., Kumar, D., Goyal, N., Verma, S. and Singh, A., 2021. An efficient framework using visual recognition for IoT based smart city surveillance. *Multimedia Tools and Applications*, pp. 1–19. https://doi.org/10.1007/s11042-020-10471-x

[13] Khan, M.A., 2020. An IoT framework for heart disease prediction based on MDCNN classifier. *IEEE Access*, 8, pp. 34717–34727.

[14] Wang, X. and Cai, S., 2020. An efficient named-data-networking-based IoT cloud framework. *IEEE Internet of Things Journal*, 7(4), pp. 3453–3461.

[15] Rathee, G., Sandhu, R., Saini, H., Sivaram, M. and Dhasarathan, V., 2020. A trust computed framework for IoT devices and fog computing environment. *Wireless Networks*, 26(4), pp. 2339–2351.

[16] Malik, K.M., Javed, A., Malik, H. and Irtaza, A., 2020. A light-weight replay detection framework for voice controlled IoT devices. *IEEE Journal of Selected Topics in Signal Processing*, 14(5), pp. 982–996.

[17] Pushpalatha, S. and Shivaprakasha, K.S., 2020. Energy-efficient communication using data aggregation and data compression techniques in wireless sensor networks: A survey. In *Advances in Communication, Signal Processing, VLSI, and Embedded Systems* (pp. 161–179). Springer, Singapore.

[18] Sheeja, R. and Sutha, J., 2020. Soft fuzzy computing to medical image compression in wireless sensor network-based tele medicine system. *Multimedia Tools and Applications*, 79(15), pp. 10215–10232.

[19] Kalaivani, S. and Tharini, C., 2020. Analysis and implementation of novel Rice Golomb coding algorithm for wireless sensor networks. *Computer Communications*, 150, pp. 463–471.

[20] Mahmood, K., Akram, W., Shafiq, A., Altaf, I., Lodhi, M.A. and Islam, S.H., 2020. An enhanced and provably secure multi-factor authentication scheme for internet-of-multimedia-things environments. *Computers & Electrical Engineering*, 88, p.106888.

[21] Bouzebiba, H. and Lehsaini, M., 2020. FreeBW-RPL: A new RPL protocol objective function for internet of multimedia things. *Wireless Personal Communications*, 112, pp. 1003–1023.

[22] Han, Y., Guo, D., Cai, W., Wang, X. and Leung, V., 2020. Virtual machine placement optimization in mobile cloud gaming through QoE-oriented resource competition. *IEEE Transactions on Cloud Computing*. pp. 1–1. https://doi.org/10.1109/TCC.2020.3002023

[23] Illahi, G.K., Gemert, T.V., Siekkinen, M., Masala, E., Oulasvirta, A. and Ylä-Jääski, A., 2020. Cloud gaming with foveated video encoding. *Transactions on Multimedia Computing Communications, and Applications (TOMM)*, 16(1), pp. 1–24.

[24] Jan, M.A., Cai, J., Gao, X.C., Khan, F., Mastorakis, S., Usman, M., Alazab, M. and Watters, P., 2020. Security and blockchain convergence with internet of multimedia things: Current trends, research challenges and future directions. *Journal of Network and Computer Applications*, 175, p. 102918. https://doi.org/10.1016/j.jnca.2020.102918

[25] Amin, S.U. and Hossain, M.S., 2020. Edge intelligence and internet of things in healthcare: A survey. *IEEE Access*, 9, 45–59. https://doi.org/10.1109/ACCESS.2020.3045115

[26] Gavrilović, N. and Mishra, A., 2021. Software architecture of the internet of things (IoT) for smart city, healthcare and agriculture: Analysis and improvement directions. *Journal of Ambient Intelligence and Humanized Computing*, 12(1), pp. 1315–1336.

[27] Rakhee, Singh, A., Mittal, M., and Kumar, A. (2020). Qualitative analysis of random forests for evaporation prediction in Indian regions. *Indian Journal of Agricultural Sciences*, 90(6), 1140–1144.

[28] Li, C. and Niu, B., 2020. Design of smart agriculture based on big data and Internet of things. *International Journal of Distributed Sensor Networks*, 16(5), p. 1550147720917065.

[29] Antony, A.P., Leith, K., Jolley, C., Lu, J. and Sweeney, D.J., 2020. A review of practice and implementation of the Internet of Things (IoT) for smallholder agriculture. *Sustainability*, 12(9), p. 3750.

[30] Singh, A., and Mittal, M. (2020, November). Prediction of solar radiation using hybrid discriminant-neural network. In *Sixth International Conference on Parallel, Distributed and Grid Computing (PDGC)* (pp. 150–153). IEEE.

[31] Aileni, R.M. and Suciu, G., 2020. IoMT: A blockchain perspective. In *Decentralised Internet of Things* (pp. 199–215). Springer, Cham.

[32] Khan, A.A., Shaikh, A.A., Cheikhrouhou, O., Laghari, A.A., Rashid, M., Shafiq, M. and Hamam, H., 2021. IMG-forensics: Multimedia-enabled information hiding investigation using convolutional neural network. *IET Image Processing*. https://doi.org/10.1049/ipr2.12272

[33] Singh, R.P., Javaid, M., Haleem, A., Vaishya, R. and Al, S., 2020. Internet of Medical Things (IoMT) for orthopaedic in COVID-19 pandemic: Roles, challenges, and applications. *Journal of Clinical Orthopaedics and Trauma*. https://doi.org/10.1016/j.jcot.2020.05.011

[34] Lv, Z., Qiao, L. and Song, H., 2020. Analysis of the security of Internet of multimedia things. ACM Transactions on Multimedia Computing, *Communications, and Applications (TOMM)*, 16(3s), pp.1–16.

[35] Liu, Y., Kong, L., Chen, G., Xu, F. and Wang, Z., 2021. Light-weight AI and IoT collaboration for surveillance video pre-processing. *Journal of Systems Architecture*, 114, p. 101934.

[36] Khan, A.A., Laghari, A.A. and Awan, S.A., 2021. Machine learning in computer vision: A review. *EAI Endorsed Transactions on Scalable Information Systems*. https://doi.org/10.4108/eai.21-4-2021.169418

[37] Alvi, S.A., Afzal, B., Shah, G.A., Atzori, L. and Mahmood, W., 2015. Internet of multimedia things: Vision and challenges. *Ad Hoc Networks*, 33, pp. 87–111.

[38] Rani, S., Ahmed, S.H., Talwar, R., Malhotra, J. and Song, H., 2017. IoMT: A reliable cross layer protocol for internet of multimedia things. *IEEE Internet of Things Journal*, 4(3), pp. 832–839.

[39] Hasan, U., Islam, M.F., Islam, M.N., Zaman, S.B., Anuva, S.T., Emu, F.I. and Zaki, T., 2020, March. Towards developing an IoT based gaming application for improving cognitive skills of autistic kids. In *Asian Conference on Intelligent Information and Database Systems* (pp. 411–423). Springer, Singapore.

[40] Ruan, J., Wang, Y., Chan, F.T.S., Hu, X., Zhao, M., Zhu, F., Shi, B., Shi, Y. and Lin, F., 2019. A life cycle framework of green IoT-based agriculture and its finance, operation, and management issues. *IEEE communications magazine*, 57(3), pp. 90–96.

[41] Nanayakkara, M., Halgamuge, M. and Syed, A., 2019. Security and privacy of internet of medical things (IoMT) based healthcare applications: A review. In *International Conference on Advances in Business Management and Information Technology* (pp. 1–18).

[42] Babu, R.G., Elangovan, K., Maurya, S. and Karthika, P., 2021. Multimedia security and privacy on real-time behavioral monitoring in machine learning IoT application using big data analytics. In *Multimedia Technologies in the Internet of Things Environment* (pp. 137–156). Springer, Singapore.

[43] Hussain, R. and Abdullah, I., 2018, August. Review of different encryption and decryption techniques used for security and privacy of IoT in different applications. In *2018 IEEE International Conference on Smart Energy Grid Engineering (SEGE)* (pp. 293–297). IEEE.

[44] Nazir, R., Ahmed, Z., Ahmad, Z., Shaikh, N., Laghari, A. and Kumar, K., 2020. Cloud computing applications: A review. *EAI Endorsed Transactions on Cloud Systems*, 6(17).

[45] Guo, R., Yang, G., Shi, H., Zhang, Y. and Zheng, D., 2021. OR-CP-ABE: An Efficient and revocable attribute-based encryption scheme in the cloud-assisted IoMT system. *IEEE Internet of Things Journal*. https://doi.org/10.1109/JIOT.2021.3055541

[46] Egala, B.S., Pradhan, A.K., Badarla, V.R. and Mohanty, S.P., 2021. Fortified-chain: A blockchain based framework for security and privacy assured internet of medical things with effective access control. *IEEE Internet of Things Journal*. https://doi.org/10.1109/JIOT.2021.3058946

[47] Kelly, J.T., Campbell, K.L., Gong, E. and Scuffham, P., 2020. The Internet of Things: Impact and implications for health care delivery. *Journal of medical Internet research*, 22(11), p. e20135.

[48] Laghari, A.A., He, H. and Channa, M.I., 2018. Measuring effect of packet re-ordering on quality of experience (QoE) in video streaming. *3D Research*, 9(3), pp. 1–11.

[49] Nauman, A., Qadri, Y.A., Amjad, M., Zikria, Y.B., Afzal, M.K. and Kim, S.W., 2020. Multimedia internet of things: A comprehensive survey. *IEEE Access*, 8, pp. 8202–8250.

[50] Huang, X., Xie, K., Leng, S., Yuan, T. and Ma, M., 2018. Improving quality of experience in multimedia internet of things leveraging machine learning on big data. *Future Generation Computer Systems*, 86, pp. 1413–1423.

[51] Laghari, A.A., Memon, K.A., Soomro, M.B., Laghari, R.A. and Kumar, V., 2020. Quality of experience (QoE) assessment of games on workstations and mobile. *Entertainment Computing*, *34*, p. 100362.

[52] Laghari, A.A., He, H., Khan, A., Kumar, N. and Kharel, R., 2018. Quality of experience framework for cloud computing (QoC). *IEEE Access*, *6*, pp. 64876–64890.

[53] Laghari, A.A., He, H., Khan, A. and Karim, S., 2018. Impact of video file format on quality of experience (QoE) of multimedia content. *3D Research*, *9*(3), pp. 1–11.

[54] Laghari, A., Khan, A.I. and Hui, H., 2019. Quality of experience (QoE) and quality of service (QoS) in UAV systems. *Multiagent and Grid Systems*. https://doi.org/10.3233/MGS-190313

[55] Bharati, S., Podder, P., Mondal, M.R.H. and Paul, P.K., 2021. Applications and Challenges of Cloud Integrated IoMT. In *Cognitive Internet of Medical Things for Smart Healthcare* (pp. 67–85). Springer, Cham.

[56] Khan, S., Abbas, N., Nasir, M., Haseeb, K., Saba, T., Rehman, A. and Mehmood, Z., 2020. Steganography-assisted secure localization of smart devices in internet of multimedia things (IoMT). *Multimedia Tools and Applications*, 80, pp. 17045–17065.

[57] Sahoo, K.S. and Puthal, D., 2020. SDN-Assisted DDoS defense framework for the internet of multimedia things. ACM transactions on multimedia computing. *Communications, and Applications (TOMM)*, 16(3s), pp. 1–18.

[58] Garg, H., Gupta, S. and Garg, B., 2021. Smart cities and the Internet of Things. *Big Data Analytics for Internet of Things*, pp. 187–195. https://doi.org/10.1109/PICMET.2015.7273174

[59] Patil, R.V., Mahalle, P.N. and Shinde, G.R., 2020. Trust score estimation for device to device communication in internet of thing using fuzzy approach. *International Journal of Information Technology*, pp. 1–11. https://doi.org/10.1007/s41870-020-00530-9

[60] Ren, Y., Zhu, F., Zhu, K., Sharma, P.K. and Wang, J., 2020. Blockchain-based trust establishment mechanism in the internet of multimedia things. *Multimedia Tools and Applications*, pp. 1–24. https://doi.org/10.1007/s11042-020-09578-y

9 Virtual Reality and Augmented Reality for Education

Awais Khan Jumani[1], Waqas Ahmed Siddique[1], Asif Ali Laghari[2], Ahad Abro[3] and Abdullah Ayub Khan[2]
[1]Department of Computer Science, ILMA University, Karachi, Sindh, Pakistan
[2]Department of Computer Science, Sindh Madressatul Islam University, Karachi, Sindh, Pakistan
[3]Department of Informatics, Ege University, Izmir, Turkey

9.1 INTRODUCTION OF VIRTUAL REALITY

Nowadays, VR is a very deep and effective technology on human perception. The VR technology importance creates a new simulated world. VR technology is becoming a special and cost-effective application and its future will have a positive impact on technology. VR delivers to us in a real world where humans can have strong and easy interface message communications. Moreover, VR technology is capable of humans as a part of new ideas inventions in the world. VR technology provides us wide concepts, which use the expanded sensory relationship with humans. VR manipulates different kinds of environments, which have special goals and enhance the capabilities in architecture, interior designers and engineers. With the help of VR, multiusers become and execute collaboration, which highlights the human's creativity, and most people can deliver their opinions. VR technology provides an excellent platform, which can stimulate the dynamic real world. As well, VR can increase productivity in different business areas and introduce a new, real world because it is difficult to imagine a graphics workstation [1]. If we look at VR history, the virtual world started in 1965. It was the first idea generated which is Real Cool, Sound Real, and delivers good realistic actions to viewers. In different decades of research on VR, research details in 1960–1962 were one multi-sensor stimulated and prerecorded film, which was in color and stereo. Then, in 1965, a second idea was demonstrated in the name of "DISPLAY ULTIMATE," based on constructing an artificial concept and reality graphic interaction sound. Figure 9.1 shows the VR architecture and its mechanisms.

DOI: 10.1201/9781003196686-9

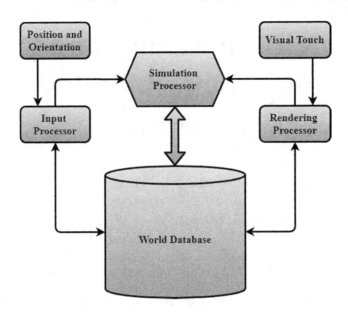

FIGURE 9.1   Architecture of Virtual Reality.

Furthermore, slowly increasing growth in VR technology moves and introduces hardware concepts, the display head-mounted. In the decade of 1975, "VIDEO PLACE," which was based on a conceptual environment, was introduced. The users can look at projected cameras on a large screen and their purpose is to define image processing and results taken from in a 2D screen space. From 1975 to 1992, different kinds of VR technology research emerged to highlight the specialty of a virtual environment and the struggle was to make relationships and interactions between instruments, applications and devices with humans. In the recent era, the introduction of AR technology rather than the virtual world emerged to replace the real world. VR is most commonly used in exchange virtual environments like adulterated experience, virtual world or artificial world. These are basic terminology. VR technology builds these three-dimensional environments as immersive, interactive and multi-sensor. Most of the business sector entered and created virtual environments, which increased the productivity and capability of humans [2].

a. **Believable:** You must believe that you are in your virtual universe (on Mars, or wherever) and maintain that belief, or the experience of VR will be lost.
b. **Interactive:** The VR environment must switch with you when you move about. You might see a 3D video and be taken to the moon or the ocean floor, but it is not immersive in any way.
c. **Computer-Generated:** What does this mean? Only powerful computers with practical 3D computer graphics will create convincing, immersive alternate worlds that evolve in real time when we walk through them.

d. **Explorable:** A VR environment must be large and complex enough to enable you to visit it. A drawing, no matter how authentic it is, depicts only one scenario from a single point of view. A book might explain a massive and complicated "new universe," but you can only explore it sequentially, just as the author explains it.

e. **Immersive:** VR must involve both the body and mind to be both realistic and engaging. War artists' paintings may include glimpses of confrontation, but they will never be able to truly express the sight, sound, smell, taste and feel of fighting. You can lose yourself in an immersive, virtual environment on your home PC for hours. The scenery can continuously shift when the plane passes over it. But, it's not the same as sitting in a live flight simulator, where you sit in a hydraulically controlled mockup of a real cockpit and sense actual forces as it tips and tilts and much less like riding a plane.

9.2 THREE HUNDRED AND SIXTY–DEGREE IMAGES AND VIDEOS

We can see the major difference between 360 degrees and 3D videos. Three hundred and sixty degrees is not a deep field, whereas 3D is a depth field but their viewpoint is limited and their technology is cheaper. Through the 360 degrees, viewers can see videos in any direction. A 360-degree video is the advancement and lower-cost technology and most of this online-based content [3]. After 20 years, the QTVR technology introduced the low-update 360-degree mobile application. Like Facebook and YouTube the first published and which had support for viewing. Thus, VLC player, Vimeo Camera 360 and 20 other brands exist in the market. A 360-degree video is a conjunction and it can be considered augmented reality, as well as more available options in new digital media technology. It depends on educators and how they convey their messages. Now the question arises whether 360 filming using 360-degree video improve the learning experience.

Advantages of 360-degree videos: Using the 360-degree videos proves the complete experience on that location, in which the viewer further engages. Presented material due to dragging 360 degrees up, down, right, and left further improves the user interaction and the user decides where and when they want to see. A 360-degree video has platforms like YouTube, Facebook and Vimeo with 360-degree video streaming and provides stereoscopic sounds. A 360-degree video directly connect viewers with content and direct alternate story. Yet, 360-degree videos provide wide opportunities, which are innovative and connect with digital media communication. It is hard to hide cuts of 360-degree images and videos and it is not possible to zoom in and out during recording. A 360-degree camera hiding the image but shows slice imperfection. Two single shots before slicing show stitching. Automation with 360-degree actions relate to direct software. A 360-degree camera is poor under low-light limitations. Virtual tours in 360 degrees: Improve presentation skills, evaluation of education interventions and research and data controller [4].

In the current era, VR and its application are rapidly making launches with famous head-mounted consumer display, rift Oculus, HTC Vive and PlayStation

VR. Recently, 3D VR applications and videos generated billions in revenue until now in 2017 and may become five times more in 2022. New 3D inventions are arising in the world of technology because VR 3D is famous and its variety of consumer applications are found in Facebook, YouTube, Netflix and most online resources. 2D and 3D VR are advancing progress technology towards the direction of 3D images and videos. Further, 3D has gathered a comprehensive database as well as acquired eye-trading data, which is based on research study [4].

9.2.1   THREE-DIMENSIONAL OBJECTS/TECHNIQUES

We can view three different directions of 3D video, which are based on reality. In the construction sector, we focus on building models dimensional, where we can observe different sides of a building or even a complete structure. Also, support an internal structure, which is convenient for both architect and customer viewing results. Secondly, the medical sector includes x-ray, medical image and further technology are an ongoing, long-term process. Three-dimensional modeling of the medical application of visual surgery in the conservation area is a very important role. 3D technology is increasingly famous in computer games, which exist as gorgeous screen effects. One better visual effect was introduced in the world of VR. 3D technology is comprehensively spread with a special impact in film production because its visual effect is fine, which shows the reality for viewers. If this is bad, then the viewer will not accept it. The updated research of 3D involves very critical issues, which need to be resolved. First, we need to improve the VR system and create a good foundation in the future. VR application systems should be simple, not complex, as well as to ensure the VR efficiency of a design system. In mind efore establishing VR 3D system based on reality which easily and simplest so that achieve targets.

9.2.2   2D/3D, VOLUMETRIC DISPLAY AND PROJECTION TECHNOLOGY

The volumetric display can provide 2D/3D images, which satisfy complete stereoscopic criteria, and vision without the need of special glasses because 2D/3D real images are made up of spotlight arranged in 3D space. The volumetric display system observed one real detect/scene that described the basic principle. The 3D volumetric display system is attracted toward the 2D image and as a result produces the principle of the volumetric display. There is high 3D image resolution through the vector scan and ray-tube monitor. 2D display device, which is an obligation, is an optical imaginary system of the 2D and 3D system based on cloned place image, which consists of three-dimensional images. CRT monitors 2D cross-section images and one scanner mirror for 3D scanning [5].

9.2.3   EXPERIMENTAL VOLUMETRIC DISPLAY IN 2D AND 3D USING AN IMAGE

2D/3D images are linked up between two mirrors: one is a cave mirror and the second is galvanometric. These mirrors are provided to help, based on 3D images with the involvement of a ector scan CRT monitor. The point drives through the scanner driver and function generator. The image supports x and y candidates and controll the brightness of images. The present computer transferring 24-bit videos/ interface 2D/3D images. For every 3D image, it is compulsory to move or switch via the cross-functional images which are displayed on the CRT monitor. The vector scan CRT monitor can do the process of the image which consists of smack lenses. Necessary volumetric display for 2D/3D: The main key factor for creating practices and effective 3D scanning method. Volumetric display is based on a rotating screen and a transferring image plane with a varifocal lens [6].

9.2.4   TYPES OF VOLUMETRIC DISPLAY

There are two types of volumetric display: one is swept surface volumetric 3D display based on the human traditional references to the optional illusion occurs when visual perception, which prepares a slice series of the 3D object into a single 3D image. 2D and 3D images prove by the persistence of vision through the human perception for continuing light of volume. It also can reflect, be transmissive or combine both; for instance, 3D scene computing and 2D screen and image. Computing based on a projection into the surface on a reply uses different kinds of techniques. In addition, changes can move and rotate. A further type of 3D display in the architecture user and member/candidate work out is the focal mirror. In a 2D screen and image based on vector, a display will compute a set of deep surfaces. Static volume: Static volume 3D displays create imagery within microscopes. Most static volume volumetric 3D encourages visible radiation into a solid, liquid, or gas. if we talk about and focus advancement implementation on static volume categories through the vapor screen or vapor display working multiple projections can result in a 3D volume space and, as a result, in a strong, static volumetric display.

9.2.5   TECHNICAL CHALLENGES

The volumetric display of most of the research claims are incomplete and not capable of reproducing scenes, which have a bad efficient viewer position. Volumetric display stands require two conditions:

1. The imagery must have consequences as a view in series as compared to slices.
2. Image background is not a spread diffuser [7].

9.3   USER EXPERIENCE OF VIDEO WITH VR DEVICES

The user experience for evaluating can beneficially use VR for most of the user experience in architecture, games and industries due to the user experience being more interactive in VR. The scientific user experience VR is more capable and efficient for viewers and users. New technology for user experience engages with industry and scientific portfolios. This area looks like an umbrella, which can be motivated between human and computer interaction because both concepts are corresponding to a good user experience. Performance generates usability and smooth user experience technology, if we consider user experience technology happens because of the interaction between a user and a product within a physical, social and cultural context. The decision designer or researcher should be aware that users have different experiences in the same direction in the product. This is the importance of users of most products in the short term. For instance, disposable things are not a strong bonding and interaction with the users [8]. The global and wide range of research of user experience and VR devices, which integrates a particular product like a mobile phone, is necessary and significant activity in a user's life. A user experiences dimensions in the following subdivisions:

    a. User researcher
    b. Human interaction
    c. Contact with the product
    d. During the interaction

9.3.1 EVALUATION OF USER METHOD

Nowadays, the need is to mainly focus on product design on visibility, which should be comfortable for users through the VR devices, because comfortable interactions check the level of accuracy. The evaluation aspects need to be achieved and observed to measure the action and performance of users. These performance parameters put the effective response described by the product. Tools and methodology analyze several human and product interaction dimensions. These tools and experiences are measured by product attributes. The emotions of the user and nature are associated with deep interaction with the product. Further, consider and focus on research-related physiological activity measures for affection responses. The main purpose of user experience defines the methods used in the evaluation and the limitations should be identified as user experiences. Suggestions strongly affected the context; thus, it would be beneficial for using VR.

9.4   CLASSIFICATION OF VIRTUAL REALITY SYSTEMS

VR frameworks are classified into three broad categories:

    a. Non-immersive VR systems,
    b. Semi-immersive VR systems, and
    c. Fully immersive VR frameworks.

Different methods of characterization include VR levels and VR tactics. Levels of VR manage the efforts needed to develop VR technology. This category includes passage stage, basic level, advanced level, vibrant structures and large-scale frameworks. Techniques for VR grouping handle the techniques used to build a VR environment. There are four types of frameworks in this category: reenactment-based frameworks, projector-based frameworks, symbol-based frameworks and job area-based frameworks.

9.4.1 Immersive Analytical and Visualization

There is immersive research on depth absorption in real and artificial VR, whereas the visualization viewer considers feelings and categorizes their decision-making steps. Visualization plays an important role in the field. VR has obtained a lot of focus and attention. Looking for a biological structure in the long term dates back to the first hype of VR technologies. The advanced immersive, analytical and visualization become part of important tools in the areas of scientific data combining visualization, data mining and different analysis methods with suitable user interactions. The present definition introduces a special issue of important topics [9].

9.4.2 Immersive Demonstrations

Immersive analytics in VR bioinformatics is based on two decisions: one is stereoscopic display and applications. These display immersive analytic creation of virtual and augmented reality systems, as well as the trend of content should be built and continue for practitioners. Find that integrated for specific visualization and interaction applications. Most of the content manuscripts presented for contribution provide a live demonstration of their implementation.

9.4.3 Immersive Analytics in VR

Important aspects of immersive analytics are normally used to support the stereoscopic visualization and enter the user-friendly environment. In addition, immersive analytics to visualization enhance the approaches in multiple senses. Immersive analytical providers investigate how to create new interactions and display technologies that can support decision making.

9.4.4 Types of Immersion VR

a. **Tactical Immersive:** Taking in process and operations for achieving results in success.
b. **Strategic Immersive:** Choose player experience and define possibilities in broad areas.
c. **Narrative Immersive:** Invested in different stories.
d. **Spatial Immersive:** Player stimulation and feelings.
e. **Physiological Immersive:** Immersion player feels confusion as in real life.
f. **Sensory Immersive:** Expiring unity a line of space focus and immersion.

9.5 3D VISUALIZATION

The 3D visualization idea combines the gathered knowledge of information sources like publications, microscope images and data information. The publications discuss collecting different information sources as well as the special arrangement for innovative stereoscopic 3D animation. The observations are relevant for creating virtual environments and must encourage visualization researchers to use 3D visualization [10].

9.6   USES OF VIRTUAL REALITY

VR has been developed in many fields essential reality areas are two which need to explain it and beneficial. Companies need to focus before designing large mechanical devices, for instance, construction sector/heavy equipment, even in the automobile sector viability and accessibility are necessary and important to the operations. Many companies invest their money manufacturing their products' interface better with the operators. Through the use of VR, companies check out the deeply comprehensive viability of their machines quickly and do some changes such as always spending money on equipment/software, etc. The second area, which VR is focused on, driving or old, used flying simulations. Flying simulations use the train track soldier virtual tank hours.

9.6.1   ADVANTAGES OF VIRTUAL REALITY

For VR live training, which is computer-based, the purpose and object increase and explore how to resolve life emergencies to improve decision making and performance. VR reduces psychological distress in a real health emergencies.

9.6.2   DISADVANTAGES OF VIRTUAL REALITY

According to the physiologist panel, the immersion virtual environment may badly affect a user, because there is no control system and it could be more addictive healthwise [11].

9.7   VR IN EDUCATION

Education is the foundation of a healthy community. From the dawn of time, civilizations have prioritized the transmission of intelligence. Educators who are constantly searching for innovative opportunities to transfer skills more efficiently, rapidly and conveniently, have used digital reality. The modern age provides an incentive to use technologies to facilitate learning. Virtual media can revolutionize the delivery of educational materials. It has now made significant progress around the world. Figure 9.2 shows the interactive model of virtual reality.

9.8   TEACHER PREPARATION AND NEW TEACHER TRAINING

Teachers are often thrown into classrooms shortly after completing their undergraduate degrees.

When it comes to bringing their skills into work, they also have a lot of real-world learning to do. Virtual reality is a means to better train teachers when they enter the classroom. Educators will learn lessons in a mixed-reality environment using programs like TeachLiVe, according to DistrictAdministration.com. Student avatars behave as though they were in a classroom during the lecture, encouraging teachers to practice their skills. This also helps current teachers by allowing them to prepare challenging lessons and assess their students' future learning. Crossing the

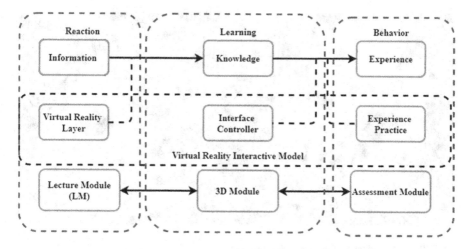

FIGURE 9.2 Shows the Virtual Reality in Education.

river According to Ed Tech, digital media has also developed a mechanism that transports educators into confrontation situations. This encourages them to practice dealing with a troublesome student or circumstance until it becomes a real-life situation.

9.8.1 DIGITAL CLASSROOM

By capturing and re-creating actual classroom sessions, VR will also assist students and teachers in gaining classroom perspectives. Teachers have also started to make accurate videos of lessons using 360-degree devices, according to tech firm VSpatial. If a student fails a lecture, they will use these VR records to virtually enter their school, see their classmates and learn as if they were there. Teachers will benefit from virtual teaching sessions as well. Teachers will learn a lot about their students' cognitive habits and their instructional methods through documenting and revising classroom sessions.

9.8.2 INCREASED LEARNING POSSIBILITIES

Virtual reality often extends the possibilities for college field trips and lab experiences. Previously, educational excursions were constrained by expense, distance and accessibility; however, virtual reality removes these obstacles and opens up a world of possibilities. From diving with whales to exploring Mars, Google Expeditions built augmented reality environments for students to discover. Students will also board actual school buses that have had their windows replaced with 4K displays to create VR landscapes. According to AR Post, science labs are being digitized to save money, have over 100 separate tests and make them more accessible to low income populations [12]. Figure 9.3 shows the advanced learning methods through virtual reality.

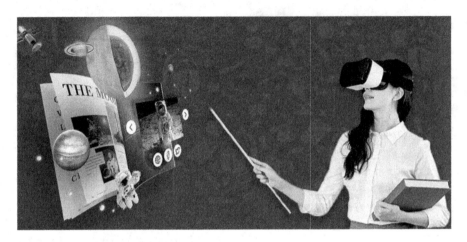

FIGURE 9.3 Advance Learning Methods [13].

9.9 VIRTUAL TOOLS FOR LEARNING IN THE REAL WORLD

Today's immersive platforms allow students to learn in a real-world environment. It ensures that students from all around the world will collaborate to solve real-world problems using virtual reality. Many colleges in various countries are working to build a more equitable climate. Language learning is one of the areas that interactive resources are being utilized. Today, students will practically study English from anywhere in the world. Digital methods may be used to build the ideal learning atmosphere no matter where students are located.

According to the Cone of Experience, people are more likely to remember what they have learned rather than what they have read, heard or seen. Virtual reality may help increase retention because people are more apt to recognize what they have encountered rather than what they have read, heard or seen. An analysis undertaken by Penn State University proved the hypothesis to be true. As opposed to students who focused on conventional programming applications, students who used interactive virtual reality were able to complete assignments even quicker.

9.9.1 EFFECTIVE ONLINE CLASSROOM

When it comes to the accomplishments of virtual reality in schooling, another noteworthy achievement is the revamped online classroom. This was largely made possible by the use of an avatar. Online classes can be made more interesting by the use of social VR apps. They offer remote students a sense of belonging. These applications may be used to extend the curricula and increase interest in online courses. It demonstrates how much augmented reality has progressed. In the past, students' failure to create memories during online classes has become a challenge. The groups may seem to be nothing more than data being fed in the most monotonous manner possible. Digital reality, on the other hand, will provide students with many ways to socialize by utilizing engaging avatars.

9.9.2 INTERACTIVE VIRTUAL FIELD TRIPS

Digital reality has made a significant contribution by allowing classrooms to participate in virtual field trips. Students at some of the country's most prestigious colleges will take virtual reality field trips and learn about new things. The best part about virtual field trips is that once they have been developed, they can be widely spread. Millions of students have had the chance to go on a virtual field trip onboard an airship thanks to Discovery Education. Google Expeditions, meanwhile, has thousands of virtual trips that help students learn by immersive media.

9.9.3 ART EDUCATION

One of augmented reality's most significant contributions has been in the field of art education. Museums all over the world have gone national in recent years. This allows art lovers to enter immersive galleries to see works from the comfort of their own homes. The Kremer Museum, for example, allows students to see 74 paintings by Flemish and Dutch Old Masters. The museum experience has been brought to augmented reality, allowing us to see paintings in the same way as in person.

9.9.4 TUTORING

Finally, augmented reality is linking tutors and students in previously unimaginable ways. Students that need more guidance will use Oculus headphones to get the support they need. They would be eligible to prep for the SAT, genetics, chemistry, algebra and other subjects as a result. Digital classrooms will be held by tutors to get students closer. They will use a virtual blackboard to sketch and do other stuff. Virtual tutoring enables students to receive the assistance they need to succeed in subjects in which they could be failing. It demonstrates how much online schooling has progressed [14].

9.9.5 CHALLENGES

The significant challenges of the VR tracking system develop better and secure. Also providing virtual systems search out more natural to allow at not the interest within a virtual environment. Most companies are working on input devices, especially for more VR applications. Most developers try to enhance productivity for a better virtual environment. While virtual reality in education has achieved a lot, there are still some issues that need to be tackled.

9.10 CREATING ADDITIONAL CONTENT

The shortage of material is one of the most significant obstacles that virtual reality faces in education. The truth is that creating more material can be very costly, and not every educational institution has the financial resources to employ a web development firm to assist them with this endeavor. It may be challenging to have startups interested in pre-undergraduate education since it does not have a lot of money. This is why it's important to enlist the help of developers and companies to help finance the creation of more material.

9.10.1 VR HEADSETS ARE AVAILABLE

While several students can afford to buy a VR headset, many students cannot. They are unable to benefit from VR-based learning as a result of this. The issue that must be addressed is giving VR headphones to all students.

9.10.2 THE SICKNESS OF THE INTERNET

Cyber-sickness is a real problem that many individuals are unaware of. It's close to motion sickness, which can make it difficult for students to concentrate. The positive news is that as technology advances, cyber-sickness is becoming less common. However, further investment is needed to ensure that students become accustomed to the sensation. To build the ideal VR classroom, educators must collaborate with businesses.

Digital reality will undeniably play a significant position in the future of schooling. It has a lot of promise, from augmented reality tools to virtual tutoring. Certain challenges, such as the need for more material, the cost and affordability of VR headphones and cyber-sickness, must be solved [15].

9.11 AUGMENTED REALITY IN EDUCATION

Some of the benefits of augmented reality in education are listed below:

a. **Accessible Learning Materials – Anytime, Anywhere:** Augmented reality can take over paper-based textbooks, physical models, banners and handbooks, etc. It provides convenient and fewer costly learning materials. This is the reason for making education much more available and easily available everywhere.

b. **No Special Equipment Is Required:** Many technologies require expensive hardware like VR. But in the case of AR, there is no need for any overpriced or costly hardware items; it is due to the reason that nowadays all teens currently own a smartphone, and the percentage is 73%. The augmented reality technologies are instantly accessible and can target a high number of markets.

c. **Higher Student Engagement and Interest:** Having an interactive and gamified nature of augmented reality provides a compelling and positive impression on the students. This helps the student to stay with the lesson until the end. It also makes the study enjoyable and uncomplicated.

d. **Improved Collaboration Capabilities:** AR applications provide a huge amount of opportunity to transform and reorganize the dull and lifeless classes. It helps in providing collective and interactive sessions, in which all the students can take part in the processes of learning and understanding at an equal and same time. This also helps in promoting skills like teamwork.

e. **A Faster and More Effective Learning Process:** Augmented reality guides the students in accomplishing their high grades with the help of forming mental visual images and full concentration on the material of subjects. Any picture can explain and say a thousand words. So, in place of reading codes and different texts about any topics, students can easily view the image with individual vision in the reality.

f. **Practical Learning:** Another advantage that we can take with the use of the augmented reality besides teaching or schooling is professional training. For instance, a perfect reproduction of infield conditions can guide in mastering the intelligence practically that is necessary for a certain career.

g. **Safe and Efficient Workplace Training:** Only with guidance, it can be done; for instance, if we are working in a dangerous environment, like managing a heart surgery in an operation room or working in a space shuttle without bringing any insecurity or having the possibility of causing damage in millions of dollars if an error occurs.

h. **Universally Applicable to Any Level of Education and Training**: Working on AR has no limits. We cannot say it is only for the specific field of use or persons. It can be used for gaming in kindergarten or for the purpose of training a trainee on the job.

i. **A Lack of Necessary Training:** Only the teachers that have the vision to look out of the box and institutions and universities with the vision to look for new technologies and new methods are ready to apply this AR application in the education system, whereas some instructors find it difficult to make it their practice due to lack of knowledge and as they do not have sufficient knowledge and skills.

j. **Dependence on Hardware:** It depends on some of the hardware and the classroom might require the systems. As we said, most of the phones owned by the students do not have compatibility with the AR application.

k. **Content Portability Issues:** The application that will be created needs to run on all types of hardware, devices and platforms. Yet it is impossible to create an application that generates equal results on all sorts of hardware and devices [16,17].

9.12 HOW AR WORKS IN EDUCATION

Augmented reality belongs to extended reality (XR) perception and theory. This theory consists of many other technologies like mixed reality and VR. AR helps in making the original world with the help of text, audio effects, images and interactive media. We can also say that this augmented reality brings an enhanced form for our natural neighboring by coating the digital content on the upper side of the visual illustration of the real world. The components used for augmented reality research are mobile phones used to play one of the most attractive games, POKEMON GO. Still, the components used in the AR glasses, for example, Google glass, dream glass or the Vizux blade, are unarguably the most acceptable and comfortable to use to hand the AR to its users. Still, the material of the AR is developed with the help of the AR software, which until now is created for the limited hardware of the AR vendors and is usually marked with the component kit. Augmented reality in the field of learning or in the field of education and discipline

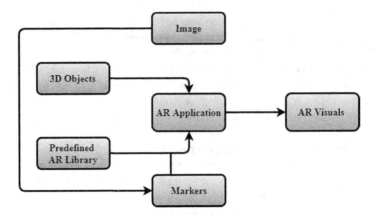

FIGURE 9.4   Augmented Reality Model.

has a large number of benefits and empowers its adopter to get the knowledge while using the environment based on the actual time directions [18]. Figure 9.4 shows the basic work of augment reality.

9.13   HOW DO WE USE AUGMENTED REALITY IN THE EDUCATION FIELD?

It is considered to be the best time at the moment to join the technology trend as we can see that the huge names in this field, like Apple and Google, are giving their full support to this augmented reality technology.

- It helps in indicating what our goal is and, according to it, who will be our audience. This selection or sorting may include the age and other functionality and, according to the app, maybe change.
- It helps us in understanding the real value after getting the idea polished by market survey and growth testing.
- It designs the POC to check the capability of the selected technology stack. It is an approved way or path for most of the new technologies like augmented and VR.
- For the testing of the foundation checking it also design the MVP; this MVP is the simple app form.
- It checks the design of the product and helps it to move towards success. It also considered that the user may require upgrading the application on a daily or regular basis with the new updates of the AR [19].

9.14   AR DEVELOPMENT IN EDUCATION: LEAD THE WAY OF INNOVATION OR ARE LEFT BEHIND

Nowadays, as we realize in the field of education, augmented reality has many roots, which are not yet discovered. Augmented reality is more easily available to use and work with the help of the current adoption of cell phone technology and with the use of another developed component. To get the most of the advantage of this market, this is the right and most crucial time to join the technology trend and to flow in the direction of the trend.

a. The AR technology grants permission to its clients to create the coat of a digital appearance and aspects on the upper side of their original natural environment and help them to display it on some sort of digital visual screen. For instance, the use of cell phones is common for the desire to show the output of the most popular game Pokémon Go. Another example is the app in the cell phone in which the phone owner records things in their neighborhood while the application changes and shows the enhanced and altered pictures having the digital appearances in it.

b. To connect the client with the technology of augmented reality, there are also different types of hardware that they can also wear other than the use of smartphones; for example, Vuzix Blade, Google Glass, Epson Moverio BT-300 and Dream Glass are some of the most known smart glasses names that have the ability of today's available AR technology.

c. Software is the basic and important tool for the reflection of output or we can say that simulation of output while the component can be anything from cell phones to the Google smart glasses. Most of the market of this software can be seen on the play store of the cell phones like the play store of the android phones or the play store of the iOS phones.

d. Yet the fact is true that some of the applications are based on specific individual hardware and are marked only with that equipment. It is also seen that the software is only designed for the use of any individual company or private use rather than for the operating system or the market [20]. Figure 9.5 shows the virtual human interaction in a lab.

FIGURE 9.5   Virtual Human Interaction [21].

9.15   AUGMENTED REALITY APPS FOR EDUCATION

We have been experiencing AR and its use around us in one or another form for decades. But the use and its importance in the field of education is reflected from the previous 1–2 years. Still in this minimum period, AR technology has shown its benefits in the process of learning and understanding. Here are some of the most outstanding examples of AR.

9.15.1   Making Learning Interesting to More Students

It is a common and known problem of the modern day's student that they are having issues in concentrating on books, blackboards or any type of handout that is used to gather information and data rather than getting it to the form on the online system. In an experiment conducted in the school of Los Angles, an experiment was conducted on the students by engaging them to use the modern-day AR system and teaching them about the solar system and the formation of the stars, which was conducted in their classrooms. This experiment became famous and beneficial from the student's point of view. In the same way, an experiment was conducted in a school in Florida, Treadway Elementary School, where AR was used to teach the students. A 3D projector was used for the experiment to show the sources of water and other features of the geography with the responsive animation responding to the movement of the children.

9.15.2   Visualizing Complex Concepts

It is true that some theories written in books are difficult to understand and are not given the proper information, whereas this problem is solved or we can say that it is made easier to understand and learn the proper knowledge with the services of AR. Forest Park Magnet School is the place where this example is proven to be true. In the experiment, the equipment used was the smart glasses, which was handed to the students and they observed the slave trade history. This was done with the help of visualization of the ship, routes through which the trade had been conducted and the documentation that was done at that time. Besides providing historical background knowledge, this wonderful technology has also served the field of mathematics. For an experiment in the math field, the elementary school in Arizona, Garden Lakes Elementary School, where the students were tested through the mobile application that intends to teach them fractions, All of the fractions were done on images and moving the fractions from here to there. This kind of fraction cannot be experienced on a piece of paper.

9.15.3   Learning Becomes Globally Accessible

Here, we know that most AR apps are created for specific reasons, whereas others are created for vast and remote uses that can be easily available to everyone and everywhere. For instance, the example that best fits is Myanmar's project. This project has the aim to make the students gain knowledge of sex education and teach them about the perfect gender relation with the help of AR textbooks. These books can be easily accessed from inside and outside of the school. The application world is widely known with the name "Human Body: Secrets of Your Life Revealed" was developed by the BBC. It is a mobile application and it has the purpose to let the user know about the vital organs. This application helps the learner to know about the heart of a human, about the liver and the brain. It also allows being tapped to provide more knowledge about different parts. It also has an integrated camera, which can be used to see the vital organs of anything.

9.15.4 REFINING PROFESSIONAL SKILLS

As we know that students of college have some skills, they can also use the AR application to make their old skills more polished and also to get them to know about the new skills. For this, the use of AR applications are already in process. Let's take the example of the Yale Nursing School, where the students are instructed that they have to apply and practice the medical practice using the ARE apps for wounds and other health problems. Some other examples may include the example of tests and assessments conducted by Dr. Amir San of UK's Imperial College of London, for which the hologram visualizing the human anatomy was shown through the headsets. This test proves to be the best way of taking the test than written the exams.

9.15.5 REDUCING HARDWARE AND EQUIPMENT COSTS

When we talk about the physical education that required the physical presence and the components, this type of training is either costly and also they are difficult for most and required lots of transport cost. With the help of the AR program, we can do any type of training in a low-cost manner. For instance, let's talk about the drill of the military, in which the cadets have to change the tires of various vehicles, but with the help of AR programs this can be done and learned in a costless way.

9.15.6 OUTLOOK FOR AUGMENTED REALITY IN SCHOOLS AND BEYOND

Having discussed several examples previously, it is now proved that the use of AR will make the education level go higher and will provide an advanced level of education and other benefits to the user. We can find numerous amounts of success stories due to augmented reality's use and this number is continuously growing. It is been clear that this is going to be an important trend in the future and all the schools and colleges have to adopt it for the betterment of education.

9.15.7 BUILDING AN AR APPLICATION IN EDUCATION

It is one of the best moves to get the niche in education and is providing many benefits. What is required? The creation of such an application may require coding skills and some specific software. If any organization has no or fewer skills regarding AR, they have to look into this matter to get the growth in the benefits or they can do this work by outsourcing it [22,23]. In the field of education, AR delivers many desired advantages, and some of them are as follows.

a   A Swift and Effective Learning System

Augmented reality helps the learner to gain the desired knowledge with the help of high-graphic images that help them in engaging and concentrating on the matters of the subject. Furthermore, the functionality of the speech technology also helps in bringing them onboard by giving them the complete specification details about the content in the form of audio [24]. Briefly, we can say that the perception of eLearning with the help of augmented reality helps the individual to spot and gather more knowledge compared to another way.

b   Easy Access to Learning Materials Anytime, Anywhere

Augmented reality can take over paper-based textbooks, physical models, banners and handbooks, etc. It provides convenient and fewer costly learning materials. This is the reason for making education much more available and easily available everywhere [25].

c   Immersive Practical Learning

Only with guidance, it can be done; for instance, take an example of when we are working in a dangerous environment like working in a space shuttle or even providing the training session for the professional like cooking food, without bringing any insecurity or having the possibility of creating damage in millions of dollars if in an error occurs.

d   Engage Students and Increase Their Interest

The gaming principle of augmented reality and the system of education, or we can say the learning system, can help the students to gain a more positive perspective. It helps in making education more amusing, enjoyable and entertaining and boosts the association and competence. Furthermore, it also gives a large range of freedom to create a class-free training with the help of providing the incomparable, attractive computer-developed surrounding [26]. The electronic resource learning gathered students in an advanced surrounding where they can experience how the theory works. For such a purpose, different organizations hire the developer deft in the development of the augmented reality [27].

e   Wrapping Up

AR technology can import the improvement to the classical learning system by bringing the revolution of the integrated schooling practices. Generally, the improvement will also encounter the passion of students and make them productive [28]. It will also help the students to get a good grip on the concepts and make the learning fun and easy to understand. With the help of offering this, the education system and environmental organizations will get more concentration and attention [29].

9.16 AUGMENTED REALITY TECHNOLOGY IN EDUCATION

All the organizations and schooling systems that are adopting and giving preference to the new and advanced technology in the classroom and also in the trip to the field are advancing to provide a new and unique way of learning to the children. In addition, with the help of AR, they are going to produce better leaders of the future. Creating an environment where students can concentrate on their path of success is the essential ingredient of the organization and it helps to move the organization towards success [30]. We are glad and thankful to the new technologies like mixed reality, augmented reality and VR that help in making learning fun, exciting, easy and fast. We can now experience many of the experiments of the different subjects from math to chemistry or biology or physics; this augmented reality helps us see the experiment from a different point of view. Now we can even show the students how the dinosaur looks or tell them about it in the class in a more fun and exciting way [31,32]. This is one way that can be said to be the best way to educate and engage the students. With the help of the applications of AR and the components, with little or even no skill of programming, makes the classroom environment is like a piece of cake. Nowadays, our physical world is expanding with the help of AR as it has layered the digital data to help us feel the digital world with our naked eyes. It changes the surroundings by including audiovisuals and images [33]. It creates a completely different and unique digital surrounding and helps us to live in that as a character or part of it like a character of Steven Spielberg's in the movie *Ready Player One*. Television and the military were the first to use the augmented reality in the 1990s when the term AR was announced [34]. But the technology has grown since then and nowadays we can see that its applications are everywhere and in every industry and have many advantages for its user as well as for the organization it is implemented [35].

9.17 OPEN RESEARCH ISSUES AND FUTURE OF VR AND AR

The VR future depends on the existing system, which addresses issues in large-scale virtual environments. In the upcoming years, according to more research, we are bound to see VR become mandatory in our homes and at work. The possible future is that we will communicate by virtual phones. Recently, the VR highlights different sectors and the results are generating billions of profit through the usage of VR (i.e., Applications, Videos). The VR is applied in different virtual environment sectors of industries and results in productivity and capabilities that enhance the use of VR. Every sector is engaged that is involved in this environment: construction, civil sector, architect, scientific and laboratory integrated with VR. The technology further enhances from VR to AR that the user environment is physically realistic and feels the visualization. The conceptual boundaries built up through VR and AR have made remarkable progress. Their most of tracking, registration solved, and their applications available in the area of education, medicine, architecture, automobile, advertising, etc. Introduce the recent real-time three-dimensional re constructor transferring in 5G communication. Technology such as artificial intelligence various human augmentation and brain science emerging and progressing with VR and AR [36–37].

9.18 CONCLUSION

VR has recently now interfered in every part of life. We cannot visualize life without the use of VR technology. In this chapter, we include the importantance and introduce the history of VR. Further, we discuss the 360-degree and 3D videos explicit in VR as well as uses of volumetric 2D/3D with projection. Furthermore, AR-based learning in education is the most important and easy-to-understand thing. Necessary of immersive and visualization for the virtual environment. We considered user experience during the 3D videos and VR devices in the VR environment. This is a fantastic opportunity for people to learn more about the new way of education. Exploration is at the heart of AR, which contributes to learning and comprehension for pupils, for both technology and educational sectors, this is a win-win situation. The rise of augmented reality is increasingly huge as applications adapt to the latest technologies. With so many benefits, AR has the potential to significantly change what it means to be human. Lastly, we have found open research issues based on the future of VR face challenges and what advantages and disadvantages maybe occur.

REFERENCES

[1] Kavanagh, S., Luxton-Reilly, A., Wuensche, B., & Plimmer, B. (2017). A systematic review of virtual reality in education. *Themes in Science and Technology Education*, 10(2), 85–119.

[2] Liu, D., Bhagat, K. K., Gao, Y., Chang, T. W., & Huang, R. (2017). The potentials and trends of virtual reality in education. In *Virtual, Augmented, and Mixed Realities in Education* (pp. 105–130). Springer, Singapore.

[3] Carruth, D. W. (2017, October). Virtual reality for education and workforce training. In *2017 15th International Conference on Emerging eLearning Technologies and Applications (ICETA)* (pp. 1–6). IEEE.

[4] Makransky, G., & Lilleholt, L. (2018). A structural equation modeling investigation of the emotional value of immersive virtual reality in education. *Educational Technology Research and Development*, 66(5), 11411164.

[5] Boyles, B. (2017). *Virtual Reality and Augmented Reality in Education*. Center For Teaching Excellence, United States Military Academy, West Point, Ny.

[6] McGovern, E., Moreira, G., & Luna-Nevarez, C. (2020). An application of virtual reality in education: Can this technology enhance the quality of students' learning experience?. *Journal of Education for Business*, 95(7), 490–496.

[7] Pinto, D., Peixoto, B., Krassmann, A., Melo, M., Cabral, L., & Bessa, M. (2019, April). Virtual reality in education: Learning a foreign language. In *World Conference on Information Systems and Technologies* (pp. 589–597). Springer, Cham.

[8] Kizilkaya, L., Vince, D., & Holmes, W. (2019, June). Design prompts for virtual reality in education. In *International Conference on Artificial Intelligence in Education* (pp. 133–137). Springer, Cham.

[9] Dergham, M., & Gilányi, A. (2019, October). Application of virtual reality in kinematics education. In *2019 10th IEEE International Conference on Cognitive Infocommunications (CogInfoCom)* (pp. 107112). IEEE.

[10] Stuchlíková, L., Kósa, A., Benko, P., & Juhász, P. (2017, October). Virtual reality vs. reality in engineering education. In *2017 15th International Conference on Emerging eLearning Technologies and Applications* (ICETA) (pp. 1–6). IEEE.

[11] Al-Azawi, R., Albadi, A., Moghaddas, R., & Westlake, J. (2019, April). Exploring the potential of using augmented reality and virtual reality for STEM education. In *International Workshop on Learning Technology for Education in Cloud* (pp. 36–44). Springer, Cham.

[12] Thompson-Butel, A. G., Shiner, C. T., McGhee, J., Bailey, B. J., Bou-Haidar, P., McCorriston, M., & Faux, S. G. (2019). The role of personalized virtual reality in education for patients post stroke—a qualitative case series. *Journal of Stroke and Cerebrovascular Diseases*, 28(2), 450–457.

[13] Gupta, J. (2019). Virtual reality in education: How schools are opting the advance retaining and learning techniques. Retrieved from: https://www.quytech.com/blog/virtual-reality-education/

[14] Martin, J., Bohuslava, J., & Igor, H. (2018, September). Augmented reality in education 4.0. In *2018 IEEE 13th International Scientific and Technical Conference on Computer Sciences and Information Technologies (CSIT)* (Vol. 1, pp. 231–236). IEEE.

[15] Chavez, B., & Bayona, S. (2018, March). Virtual reality in the learning process. In *World Conference on Information Systems and Technologies* (pp. 1345–1356). Springer, Cham.

[16] Chen, P., Liu, X., Cheng, W., & Huang, R. (2017). A review of using augmented reality in education from 2011 to 2016. *Innovations in Smart Learning*, 13–18.

[17] Akçayır, M., & Akçayır, G. (2017). Advantages and challenges associated with augmented reality for education: A systematic review of the literature. *Educational Research Review*, 20, 1–11.

[18] Hantono, B. S., Nugroho, L. E., & Santosa, P. I. (2018, July). Meta-review of augmented reality in education. In *2018 10th international conference on information technology and electrical engineering (ICITEE)* (pp. 312–315). IEEE.

[19] Khan, T., Johnston, K., & Ophoff, J. (2019). The impact of an augmented reality application on learning motivation of students. *Advances in Human-Computer Interaction*, https://doi.org/10.1155/2019/7208494

[20] Garzón, J., Pavón, J., & Baldiris, S. (2019). Systematic review and meta-analysis of augmented reality in educational settings. *Virtual Reality*, 23(4), 447–459.

[21] Alex Shashkevich. (2019). New Stanford research examines how augmented reality affects people's behavior. https://news.stanford.edu/2019/05/14/augmented-reality-affects-peoples-behavior-real-world/

[22] Soltani, P., & Morice, A. H. (2020). Augmented reality tools for sports education and training. *Computers & Education*, 155, 103923.

[23] Yip, J., Wong, S. H., Yick, K. L., Chan, K., & Wong, K. H. (2019). Improving quality of teaching and learning in classes by using augmented reality video. *Computers & Education*, 128, 88–101.

[24] Yilmaz, R. M. (2016). Educational magic toys developed with augmented reality technology for early childhood education. *Computers in human behavior*, 54, 240–248.

[25] Ling, H. (2017). Augmented reality in reality. *IEEE MultiMedia*, 24(3), 10–15.

[26] Masmuzidin, M. Z., & Aziz, N. A. A. (2018). The current trends of augmented reality in early childhood education. *The International Journal of Multimedia & Its Applications (IJMA)*, 10(6), 47.

[27] Huang, T. K., Yang, C. H., Hsieh, Y. H., Wang, J. C., & Hung, C. C. (2018). Augmented reality (AR) and virtual reality (VR) applied in dentistry. *The Kaohsiung Journal of Medical Sciences*, 34(4), 243–248.

[28] Lin, C. Y., Chai, H. C., Wang, J. Y., Chen, C. J., Liu, Y. H., Chen, C. W.,... & Huang, Y. M. (2016). Augmented reality in educational activities for children with disabilities. *Displays*, 42, 51–54.

[29] Klimova, A., Bilyatdinova, A., & Karsakov, A. (2018). Existing teaching practices in augmented reality. *Procedia Computer Science*, 136, 5–15.

[30] Koca, B. A., Çubukçu, B., & Yüzgeç, U. (2019, October). Augmented reality application for preschool children with unity 3D platform. In *2019 3rd International Symposium on Multidisciplinary Studies and Innovative Technologies (ISMSIT)* (pp. 1–4). IEEE.

[31] Cabero-Almenara, J., Fernández-Batanero, J. M., & Barroso-Osuna, J. (2019). Adoption of augmented reality technology by university students. *Heliyon*, 5(5), e01597.

[32] Rau, P. L. P., Zheng, J., Guo, Z., & Li, J. (2018). Speed reading on virtual reality and augmented reality. *Computers & Education*, 125, 240–245.

[33] Vávra, P., Roman, J., Zonča, P., Ihnát, P., Němec, M., Kumar, J.,… & El-Gendi, A. (2017). Recent development of augmented reality in surgery: a review. *Journal of Healthcare Engineering*. https://doi.org/10.1155/2017/4574172

[34] Sungkur, R. K., Panchoo, A., & Bhoyroo, N. K. (2016). Augmented reality, the future of contextual mobile learning. *Interactive Technology and Smart Education*, 13(2). https://doi.org/10.1108/ITSE-07-2015-0017

[35] Turkan, Y., Radkowski, R., Karabulut-Ilgu, A., Behzadan, A. H., & Chen, A. (2017). Mobile augmented reality for teaching structural analysis. *Advanced Engineering Informatics*, 34, 90–100.

[36] Laghari, A. A., Jumani, A. K., Kumar, K., & Chhajro, M. A. (2021). Systematic analysis of virtual reality & augmented reality. *International Journal of Information Engineering & Electronic Business*, 13(1). https://doi.org/10.5815/ijieeb.2021.01.04

[37] Jo, D., & Kim, G. J. (2019). AR enabled IoT for a smart and interactive environment: A survey and future directions. *Sensors*, 19(19), 4330.

[38] Alex Shashkevich. (2019). New Stanford research examines how augmented reality affects people's behavior. https://news.stanford.edu/2019/05/14/augmented-reality-affects-peoples-behavior-real-world/

10 Fog- or Edge-Based Multimedia Data Computing and Storage Policies

Preety[1], Kuldeep Singh Kaswan[2] and Jagjit Singh Dhatterwal[3]
[1]P.D.M University, Bahadurgarh, India
[2]Galgotia University, Noida, India
[3]P.D.M University, Bahadurgarh, India

10.1 INTRODUCTION

In the research, designing, scripting and development of hardware, software and connectivity facilities, multinational corporations pay huge sums on wages for technology professionals and there are direct results or inventions that are incorporated in real structures in the vision of the Internet of Things (IoT). IoT systems have been generally misunderstood to be exclusive to sensors and the detection of radio frequencies. Surely this is not a definition of IoT. Comprehensive technology exists among Bluetooth, Wi-Fi, sensor systems, electrical systems, social networking sites, 4G/5G, wireless communication, EPC, barcodes, IPv4/6 and QR codes. The IoT framework consists of AI technology, robots, environmental knowledge, websites 3.0, large data storage and mobile technology data analytics processing, etc. [1,2]. Indeed, all current applications include Window frames 16, Linux, Microsoft Windows, SuSE Linux, Apple computer, Ubuntu, Android and Sun Solaris. OP structures are further extended with IoT technologies. In reality, the IoT world is the bedrock of electronic hardware equipment, including computers, notebooks and servers. These innovations allow people to comfortably live and change public services. To make them more fit, building automation, academic efficiency, sound governance, digital engineering, defense facilities and health care – all in a big way. Any of these innovations currently exist and are publicly available, except for technologies in the enhanced form. Also, items such as desks, chairs, cabinets, walls, refrigerators, washes, clothing and glasses will be linked through that same development to the future Internet (FI). Even ranges, lakes and woodlands can be linked and tracked and secured from deliberate damage by Wi-Fi or wireless local access. Indeed, inanimate natural or man-made structures will communicate with *Homo sapiens*! The new technology for IoT is proactively being developed in Europe, Singapore, Korea, Japan, China and elsewhere in South Asia

DOI: 10.1201/9781003196686-10

to enhance market productivity and change human lives daily. Various joint efforts are underway in developing devices for electronics, applications and networking. Many researchers use IoT architectures, interoperability and protection protocols when transmitting data between IoT devices[2]. Many foreign companies, companies, associations and universities have agreed to produce IoT products in the IPSO Alliance. A further important consideration is the protection of computers, data transfer and the development of higher-quality technologies and mechanisms for many global business giants. For the protection of personal records, now international cyber regulations are being implemented. Different objects linked to the Internet produce a large volume of information. However, various disadvantages are computer problems, shortage of storage, lack of speed and safety issues are just some of the emerging technology challenges. A modern technique called "fog computing" is steadily taking its place on the market to address these problems. Widely adopting the fog computing model, it is important to build modern network architectures and middleware solutions that deploy new, evolving communications technology such as 5G, advanced IoT analysis software, big data and deep learning for a new context-sensitive approach. Fog computing can enable the computation of devices that produce and need transmitted data at the edge devices immediately. Lack of speed and safety issues are just some of the emerging technology challenges.

10.2   INNOVATION SERVICES

Fog computing is an extreme concept that is often used to replace cloud technology, although variations occur, as is seen in the following debate. Fog intelligence is a recent methodology that further expands the framework of cloud technology where data-generating devices are located. Fogging offers end-user computing, data management and smartphone services. The innovations and services [3] are facilitated. When the IoT hypothesis is realized, a huge amount of computers are connected to the Internet of the Future (FI). Each tangible process and item connecting to the digital infrastructure generates large volumes of data each couple of seconds, in giga, tera, peta, exa and zetta, among other circles possibly even in yottabytes. Therefore, the current cloud computing infrastructure system will not be able to carry out the necessary analysis at the time necessary. Because of the large data collected by connected heterogeneously variable devices and high transmission speeds, the data transmission capacity does not have the necessary caliber to handle all of this data on one occasion. Additionally, inappropriate data transfers to the unauthorized wide area networks, state zone networks, there will also be the processing of networks for urban areas, municipal channels and home areas.

10.2.1   PRINCIPLE OF COMPUTING

The principle of fog computing is rescued in this case. Cloud computing has emerged as an alternative to the hardware and service paradigm of information technology. However, the cloud fails to address local problems affecting a vast range of networked components effectively and, as for all technically unified technology networks and services, it is not sufficiently flexible for certain applications that need the direct intervention of a local monitor [4]. The standard cloud infrastructures, as mentioned previously, are not sufficient to meet the increasing needs of IoT applications. The latency and low network constraints are two big problems. The research [5] reveals that the fog computing or edge computing ideas have been suggested in recent years to boost these shortcomings by bringing transaction processing capacities nearer to the network edge.

10.2.2 Meaning of Edge and Fog Computing

The concept of fog computing is sometimes referred to as "Cisco" in networking. The product line manager, Mr. Ginny Nichols for Cisco, is thought to have coincided with the concept and was hence credited with fog computing's ancestor. Although the name is "Cisco fog computingComputing," fog computing technology is documented and accessible worldwide. Cisco recently gave a fog computing perspective, which allows applications to specifically run in the network on trillions of wired machines. On the computer networks' popular packaging from Cisco, including hard drives and adapters, customers can build, manage and operate software applications. Leaders from Cisco, Dell, Intel, Microsoft, ARM and Princeton University formed in November 2015 the Open Fog Consortium. The key goal is to create the open database images and communicate fog computing's commercial advantage to the "Open Fog Consortium." The next limit for speeding up IoT communications is fog computing, providing pace and consistency, and many other advantages. Fog computing is a heavily virtualized infrastructure, usually situated on the edge of the network, providing computing, storage and networking capabilities between terminal devices and conventional cloud computing data centre. Fog computing provides end users with data, computing, storage and application resources. Some people say that "fog computing" is coined as "fogging." Prof. Salvatore Stolfo. Fog computing is described as a communicated cloud computing that handles the applications and resources either on the network or in the cloud of the distributed data center. Therefore, the "switches," the "mugs" connectivity and the connected devices become measuring appliances for local computing in picoseconds. Cisco [6] proposed the idea of fog computing in a comparison with computing in the cloud and the storms are farther from the earth, but the fog smaller. The word fog computing is thus assumed to emulate closer to target consumers the minimal cloud technique. Fog computing is specifically applied to encourage acceleration implementations and a broader perspective. Everything is always Internet (IoE) technology [6] has been envisaged as innovations for mitigating real time, mobility and knowledge of the place, such as fog, mobile edge, clouds and micro data centres. Fog computing is available to the whole world. When choosing the term "fog," the intention is to make the benefits of cloud storage close to the database. "Fog" is just a club near the field of atmospheric science. In an evolution that was marked as "The Fog" [7], the cloud migrates to the data centre where it will become the application development infrastructure itself. Any sustainable home should have its mobile phone or intelligent landowners' fog electronic equipment tablets using this technology.

Fog computing can help address the needs of many emerging technologies, such as personality vehicles, traffic lights, intelligent homes, etc. than cloud computing. However, it cannot completely substitute cloud computing – since clouds are also preferred to be used for the most popular high-end batch processing tasks in the world of industry. So we should conclude that fog computing, cloud computing and their benefits and drawbacks balance each other. Fog (edge) technology plays a very important part in the field of cloud computing.

10.2.3 SCIENTIFIC EDGE AND CLOUD COMPUTING

Essential studies in the cloud computing infrastructure include study results in connection with anonymity, safety, device stability and safety mechanisms. Fog computing will expand to support the new network concepts that will need quicker cloud computing helps businesses to meet their increased processing compliance and address their expenses while creating fewer complications by price and feed. The stability of intelligent device networks. It follows that the concept of a stable fog platform is a workable model combining the performance specifications of the grid and cloud frames with transmitting and electronic controls accuracy and stability. The fog computing principle can also be regarded as a surface extension [8]. Instead of developing online infrastructure and networks, other types of transfers and facilities are located on the bottom of a communication network. Exhibitors of the fog computing discussion are that it will reduce any need for capacity simply by not transmitting all the data over the Internet. Collect it at some points of entry, like modern routers. This computation enables a more effective information compilation which, if at all, is not immediately required in the cloud. In this type of dispersed approach, the project managers of the organization will guarantee reduced costs, faster operation and increased system efficiency. In a connected computer like a smart or Android mobile device, a smart firewall or a bugger or networking equipment processes in the fog computing world of the data center. This limits the volume of data sent over a network to the cloud. In particular, it must be remembered that fog network additions do not at any time overtake cloud storage; fogging enables short-term analysis to be carried out at the front, while cloud computing technology carries out long-term analytics with great intensity of resources. At first, when the data are generated and compiled by edge sensors and devices, they have no computing and storage capacity to perform sophisticated analysis and modeling activities. Though database servers can, they are much too far away from storing and responding fast. In comparison, the gathered information is connected and shared over the network to the database for any single endpoint can have confidentiality, safety and serious consequences, especially in the context of sensitive data subjects in various countries. Fogging is the safest strategy in this situation.

The benefits of fog computing in several areas, including wireless sense networks, the smart grid, things' Internet and established software networks are many. However, the migration of anonymity, trust and security services between fog computers is general questions associated with fog computing. By 2023, the global fog market is forecasted at $556.72 billion and expected to grow in the forecast period 2017–2023 with a CAGR of about 61.63%.

10.3 FEATURES OF FOG AND EDGE COMPUTING

Fog computing is enabling a broad variety of technology and services [6] to extend the idea of cloud technology to the top of the loop. Features of edge computing include low-power consumption and position sensitivity:

- Widespread regional deployment
- Movability
- A broad several of network nodes
- Prevailing position for wireless connectivity

Data reliability fog computing was described by several authors: The key characteristic is that fog computing serves boundary conditions of the finest networks at the edge networks with low latency and position knowledge. Secondly, it has become very desirable in the geographical distribution because facilities and applications are widespread around countries and spaces in the fog. Strongly opposed to the more unified cloud, fog computing technologies and frameworks need widespread implementations. For example, fog can provide appropriate high-quality transmission of moving cars utilizing equivalents and roads [9]. The third feature is accessibility assistance, i.e., the use of LISP protocol fog to include movement strategies such as host identity disconnection. That means that it is important to connect various fog and edge computing uses directly using smartphones and other intelligent or computer equipment. Accessibility strategy assistance, for example the LISP method, is also essential to decouple host identification from the location and to provide a distributed directory structure.

- The ability to communicate in real time in a different trait and a key prerequisite. Fog computing involves real-time communications without disruptions for fast service.
- The bandwidth height for heterogeneous network applications. In a wide range of conditions, sensor nodes may be implemented. These worlds contain items or machines that are of heterogeneity, such as computers, servers, notebooks, scanners, pagers and smartphones and paying objects, both living or not. So you can envision a house where all – from equipment topets – WLAN.
- Accessibility is an important feature of the fog computing network. This illustrates the interaction and cooperation of multiple hardware devices or IoT artifacts over various networks during the data transmission process. It includes a broad spectrum of features including framework downloading and fog components must be capable of interacting.
- The aim is to increase the general public's quality of life. This means overcoming bandwidth constraints in this homogenous setting, increasing processing capacities and processing and fogged up is the best strategy for providing the necessary resources.

10.4 CONCEPTS OF FOG COMPUTING

Fog information technology offers a range of storage and communications benefits.

> the staffing levels can be nearly exponentially improved by bringing data processing
> and supply closer to the requested devices by expanding the regional network footprint
> and running in cluster head fog infrastructure than either in the cloud or in conven-
> tional systems.

The consequence of data flowing to the user along a transmission network route across such fog nodes simply cannot be overestimated. Centrality is the main contributing factor to network performance. The increased effectiveness of the network through proximity increases some of the classical design benefits by reducing their inconveniences. Network congestion is also dramatically minimized through this transfer to network nodes. The overall processing in a much-unified framework is needed in the conventional architectural model. In all but a few special cases, traditional architectures are glamorized bottlenecks, where data are stuck in the pattern of a looping hold. By the distribution of this functionality through a much broader space, fog computing works. Although such technologies can deny cloud distribution, this is troublesome, since it also ensures that, along with multiplexing and other packet filtering solutions, the whole network response suffers. Users can also face inefficiencies, but the load balancing principle has been used extensively to deliver faster traffic and increase network reliability. With the addition of nodes, you can spin and use dormant nodes. Fog computing has excellent skills in scaling. Working with nodes means that the data demand is geographically close. This also signals an increase in reliability, as data is encoded as it travels to the network edge; and a change in node relationships and existence ensures that the surface of the user attack is still improving. Fog and cloud have three layers of service delivery that serve a range of applications, including that of the production of online content [10], rising reality [11] and predictive analytics [12]. However, there are several issues with fog/edge computing too. It improves the transmission lines and introduces a certain amount of overhead – both globally and worldwide. The only disadvantage of fog technology and architecture abroad – in business terms. In addition, the implementation of edge computing ensures that an increasing number of failure points are being introduced by the customer. Although this continual shift in privacy and protection benefits, the facets of maintenance and likelihood are not. In addition, during the construction of the fog computing scheme, protection and privacy should be discussed in each layer [13,14]. In the past, the customer has one single failing point for a consolidated repair, maintenance and identification initiative to detect possible problems within the fog computing architecture. Users spread this initiative and obligation by distributing the technical burden, which can pressure the whole fog computing operation. Fog is another virtual network environment layer that is loosely related to technology and the Internet. The edge position, location awakening and reduced power mean fog computing serves data sources at the network's edge with the most advanced services.

10.5 MODELS/ARCHITECTURES

No typical frameworks and prototypes validated for scientific testing were derived. However, in the subsequent subsections, some of the fog computing architectures were listed to enhance the reliability of the fog perception.

10.5.1 A GENERIC FOG COMPUTING ARCHITECTURE

A general fog computing framework consists of the following elements developed by the Mind Entrepreneurship [15]:

1. IoT endpoints including final sensors, access points, fasteners and other processing devices that execute the necessary tasks, as well as applications built on intelligent artifacts.
2. There are IoT endpoints. This device is generally linked to the storing of records. Data from other users are provided for the IoT endpoints. Some data are stored in real environments in the IP network information processing units. Observation two was stored in a cloud infrastructure for further use.
3. Network IP (Internet control message protocol) – provides data for collecting and eventually transmission of required data from and to the cloud world and attached to application IoT devices (e.g., transmitter). The main objective is cooperation. The IP network uses sensing technologies to assist the dissemination of knowledge such as massive data mining, etc. Distributed intelligence is seen to provide an important cognitive foundation for the understanding and assessment of human accomplishments, objects, tools and socio-technical environments, which enable living creatures. The knowledge is collective intelligence.
4. Centralized unit for control and intervention – a cloud-based computing device that, if necessary, saves and investigates all data not accessible by a software development device. For business forecasting, research and investigations, the computer is used.

10.5.2 FRAMEWORK FOG AND EDGE COMPUTING MODEL

The whole fog computing framework has been separated into three layers (Figure 10.1):

- **Centralized Cloud Services Knowledge layer:** The network infrastructure platform could be used here for the collection and management of databases. For key communications and telecommunications, it is important to provide an Internet (IP), multiple label switching (MPLS), operating performance, multicasting, information assurance and networking solutions.
- **Fog Computing: Distributed Intelligence layer:** This layer is the key layer on the integration node. You may also change the name of the low- and mid-edge. 3G/4G/Long in this surface logical controller (PLC) technologies, Wi-Fi, Ethernet and other such technologies can be found. The ground area network is just like that.

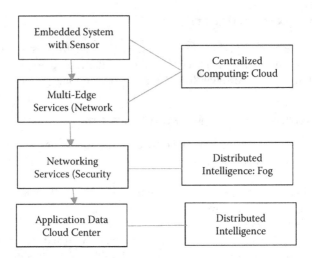

FIGURE 10.1   Framework Fog and Edge Computing Model.

- **Distributed Intelligence: End-Point Computing layer:** The smart things network can also be renamed. There are many inserting structures, artifacts and detectors in this layer. This shows that intelligent and less intelligent things such as vehicles and equipment are linked by wireless connectivity links.

10.5.3 A Fog Computing Architecture

The fog architectures have been divided into five layers:

1. **Data management layer:** IoT sensors and devices constitute the bottom layer of technology. This layer is dispersed globally and has detecting, communications and implementation aims that guide the values collected to the next layers via gates for further transmission. In this layer, the IoT actuators contribute significantly in regulating the process or IoT structure. The data was collected from the remote sensors and converted into electromechanical periods. The actuator normally answers detectors that feel the difference in society.

2. **Intermediate result computing layer:** The next step is "fog computers," gates for network detector content, remote controller and mechanical computers. The layer will host the various kinds of program modules. Generation of data in this layer can process the data obtained from various sensor nodes, RFID and other Internet of Things; the remaining can be forwarded to a cloud environment for an efficient manufacturing process. This layer produces a huge amount of information.

3. **Edge computing layer:** The method monitors the use of resources, prerequisites for cameras, electronic controls and IP accessibility. It monitors and records the tools and implementations of technologies. The tracking

aspect provides information collected for other programs and application purposes, depending upon the requirements.

4. **Sensors network layer:** This layer basically manages data collected from different sensor nodes and its task is performed by different nodes. Resource processing is the central fog's design sheet. It is made up of computer modules that systemically control resources to minimize resource waste Null. This layer is designed consistently and computerized to manage the smooth control of resources. It offers common application process interfaces (APIs) for hardware components such as processor, storage, computer, IoT computers, etc. to manage and control. The positioning and scheduling modules are the main parts of this layer since they monitor the status of the available resources.

5. **Physical layer:** This fog layer consists of an intellectual procedure model in which the device releases measured data on a daily or final basis. Fog-friendly systems attach data from the sensors and process it.

Finally, the management perspectives are conveyed to the actuators. There are brief descriptions of two related models below:

• Model sensory processes: The gathered input is transmitted as data sources, acting on software operating on fog devices, and the resulting controls are transferred to transducers.

• Streaming data model: The parallel programming architecture has an application network operating on fog networks and concurrently manipulating sensing connected devices. In wide and long-term analysis, information obtained from incoming sources is saved in data centers. We regard load balancing as a separate category of the sensory-process model [16]. The Internet of Things (IoT) implementations left no fields, social or trade, unaffected. IoT possibilities involve health care, manufacturing, management, housing development, home automation, protection for people, schooling, transportation, defense and much more [17].

10.5.4 FOG COMPUTING TREE MODEL

In the development of a model of the Internet of Things (IoT) ecosystem, the "tree model" was named. It defines the three-layer performance bottleneck. It was built on IoT Forum Architecture, ITU Architectural style and IoT Architecture Templates of the European FP7 Research Project. This model is supplemented in the FP7 research project by IoT-ARM (Architectural Reference Model). This method explains the modules regarding their communication and process movement. IoT's "forest model" has roots, trunk and infinite leaves as a tree system [18–20]. The processing, transport and application layers as explained below are the three components of the tree model.

• **Processing layer:** The roots are the foundation of a tree and are important for the development of the branch and its whole existence. Similarly, the computing layer is the main layer with fully integrated capacities for the environment of IoT in an SDN infrastructure. This layer usually can compute and integrate various physical structures or items. The technology concerned supports the identifying, protecting and tracking of items. These systems are naturally even capable of automating the network communication protocols, which can eventually influence even people. Any of the processor technology included smart cards and RAI, CT, detectors, electric motors, security code, IPv4/6, QR codes, NFC, Wi-Fi, wireless and ZigBee Speech 5 High energy mark-ups, robotics, microprocessors, advanced analytics, UOS, browsers, Linux, smartphones and networking websites.

- **Network layer:** There are several feature levels and synchronization located between the network topology. There are, therefore, no clear guidelines to repair the sequence, and edge computing will possibly merge the processor layer and session layer into one another. The layer of the network consists of a Home Automation Network, LAN, MAN, SAN, WAN and alternating subsections. HAN is a new and relatively unidentified network. Innovations allow for us to connect to the Internet, phones, clothing, cookware, curtains, doors, desks, fans, coolers, etc. This Internet protocol offers high bandwidth and high data byte transmission packets for improved utilization efficiency. Digital 5G, including NG9-1-1 engineering, will be in operation shortly to improve more. Fog computing can happen through this network layer controller, thermostats, electric motors and intelligent portable devices.

- **Application layer:** The layer of application is analogous to the "tree" leaves. The IoT application layer contains a range of technologies like transportation, safety, administration, learning, health care, pharmaceutical, logistical, industrial and process industries, as do countless leaves on the tree. In addition, the application layers promote a better quality of life technology for public utilities in metropolitan environments. Services involve electricity management, water treatment, waste management and irrigation. Indeed, this also enables mining activities to be tracked and natural habitats, including swamps, forests, lakes and mountainous areas can also be controlled. IoT has provided a huge range of quality of life programs 24 hours a day.

10.6 COMPARISON BETWEEN CLOUD COMPUTING AND FOG COMPUTING

Like cloud computing, the contours of fog computing must be understood first, to understand what cloud computing is all about. A thorough understanding of the rational and philosophical expenditure of cloud technology is necessary before it is understood. The idea is to express the distinctions between the two definitions and explain why such a demarcation is essential.

Internet and cloud computing are also philosophically dependent on the Internet of Things (IoT). Technological innovators have constantly promoted the use of cloud computing to increase IoT implementations. The IoT expresses an orbital space network that provides access to various services, which are based on either a network of individuals (802.15) or the local network (802.11); a network of the metropolis (802.16); and a network of broad areas (802.16). (802.16). The basic components required for the design of complex and competitive applications in both cloud and fog computing environments are required. Below you can analyze each of these technical blocks:

Architecture Service-Oriented (SOA): the repository contains a test and feature-oriented applications library which can be modified to be useful.
Language of Extensible markup (XML): allows for the use of tags to carry the knowledge of any type through any specified Internet-based program.
APIs: technological labels for use of relevant Internet-related web applications.

Edge computing is an environment of the machine that combines stores, processes and communicates with the edge computing consortium cloud-to-thing [21] to get end users connected [21]. Fog computing uses edge computing to push virtualization to another level. Virtualization allows open network elements to be used to create virtual machines or states. Cloud computing mainly focuses on providing information across an almost linked central network, while cloud technology focuses on the provision of distributed documentation to the Internet of Things to serve new and evolving applications focused on technologies. A leading-edge router with constructed network-attached storage, speed of information processing and the like is an example of advanced technology [4]. Cloud technology is a cloud computer addition including channel differences architecture in which the network edge is used to execute the network edge and the network core is used [4]. Fog computing is commonly understood to interface with interdependent and unseparated ubiquitous systems and to collaborate in a situation where such computers perform storage and processing tasks. Performance characteristics are used for fog computing. In case fog computing is used to analyze and transform information at a repository stage, it reduces data delay dramatically and enables smooth data delivery. In this respect, due to its client system and local context features, the fog server enables complex, adaptable optimization. The fog server will provide more experience for users to the information used to enhance the rendering of web pages and surf experience.

Instead of fog computing, applications are used such as computer paradigms like Mobile Cloud Computing (MCC) and Mobile Edge Computing (MECs). A cloud service at the end of a mobile network may be an example of the MEC application, where functions could not historically be performed on a convenience network [4]. Any of the common products developed in the fog computing conceptualization [4] are the Google Mirror, Sony Smart Eyeglass and Microsoft HoloLens. In the phase of development where data is processing massive data collections, fog computing allows the possibility of improved capabilities. This processing allows the collection, aggregation and preprocessing of large amounts of data to allow data to be transported and thus to balance computing resources. In addition, fog computing expands the advantages that the Internet of Things will draw.

10.7 PROMISE OF CLOUD AND FOG COMPUTING

Specialists unquestionably agreed that the introduction of the cloud and cloud technology harms accessibility and ease. Fog computing has little obsessive focus on the ability to change the way finished computing is viewed. There is, however, also a study that has shown that cloud or cloud technology does not always lead to profits and cost reductions when, in some situations, the profits can be less than those for the whole model. In addition, in some contextual environments, the "economic value" associated with cloud computing is often strong. In particularly resource-restricted environments, such as Africa, the idea of opportunity costs begins when assessing whether fog computing should be considered. The unexpected high costs of cloud computing can in the longer term be very expensive for the business by transferring huge data sets to the cloud and saving them there. Moreover, moving the large data sets in between the business and the cloud will lead to massive bandwidth costs, particularly if there is poor throughput and large data duration. Cloud and desktop virtualization offers the following main advantages:

- The ability for big companies to transfer part of their positions in data management, translating any of its target areas beyond high beginning costs to the cloud and small businesses, especially beginners.
- Enables device developers to concentrate on the core logical design of the industry without needing to consider maintaining the hardware and expandability.
- Service users can use facilities that are all too comfortable.

10.7.1 FUNDAMENTAL CONCEPT IN EDGE, CLOUD AND FOG COMPUTING

Scholars and clinicians sought to describe the basic principles that control the design and performance of the Internet of Things and cloud computing. The open geospatial proposed framework encapsulates the ideas that, for example, govern the storage, network, computer, control and accelerate configuration [21]. To construct fog and grid computing implementations, the following concepts are fundamental:

- The ideals of durability, accessibility and maintainability should be respected in prototypes. The standard of cloud providers should still be available to process transactions. The need for uniform usability in fog applications, however, opens up various safety problems and difficulties in these applications. Real-world fog implementations enable the security laws to be exposed to various security aspects.
- Tactical and practical decision making offers agile applications cardinal to data and knowledge conversion. Whenever there is a justifiable shift in customer's expectations, cloud systems can be so agile that they make improvements in customer settings and service models.
- Computing systems fully virtualized to observe independence at all concentrations, large proportions and virtualization. Cloud platforms can be

made available to end users. Around the same time, the efficiency of services does not decrease system security and independence, and real data value observation.

- Implementation requires specific and respect for the desired privacy values. Cloud platforms can have a greater degree of scalability because technologies evolve, and new services and apps arise each day.
- Fog computing understands the layout of the fog network node. The shortest fog server is an analytical device that can be used to consider the fog framework.
- When planning fog networks in the network layer, the different fog computing elements must be dealt with. Cloud technology includes:
- Online Analytical Processing (HDFS), Inventory monitoring Hadoop [7], Dryad [22], virtual machines and the hierarchical motor for the use of ragged data [8] – light in weight runtime model [10]. The cloud computing technology includes: In parallel computation, these developments have opened up exciting possibilities, particularly for managing massive data systems [11].
- Manjrasoft Aneka offers itself a forum for developing modular applications in various clouds in a streamlined and elastic way [11], infrastructure projects like the free software club Nimbus [12] and Melaleuca [13].
- Cloud infrastructure devices like the cloudlet, Cisco IOx, are among the commercial modern technology.

10.7.2 TAXONOMY OF EDGE COMPUTING

- NIST Cloud Cloud Design Methodology is a conceptual high-level paradigm that focuses on what cloud services are required, regardless of the design process. It offers a controlled cloud services terminology by expressing the cloud taxonomy computer. No individual cloud computer or technology product is linked in the structure. The cloud carrier is linked to the cloud vendor, the cloud trader, the cloud username and the cloud auditor in the logical proposed methodology. Each aspect of the reference model has certain characteristics that have to be taken into account when designing cloud or grid implementations.
- The security audit, safety audit and efficiency audit are available to the cloud auditor. There are separate audit schemes that ensure transparency in the provision of over-cloud applications. By describing the physical infrastructure that contains the hardware and the facility, the cloud service providers work on service orchestration. The operating layer takes cloud computing's diverse service structures like the platform as a service into account. The orchestration of the operation is usually the resource abstraction and control layer.
- The cloud process improvement component includes maintenance, supply, setup and the functionality or compatibility of cloud and grid systems platforms. The cloud broker was responsible for the financial intermediary, consolidation and arbitration of services. Technologies, such as a chip

device(s), are capable of incorporating the entire system to a single unit
and widely used in smartphones (i.e., processor, memory, timers, communication protocols) and a system in packages (SiP) can be used for fog
computing [13].

- Web 2.0 apps serve to interactively and flexibly share content, build user-
centered and collaborative and compose applications offering improved
user interfaces. This is possible thanks to the use of innovative technologies like XML, web servers and JavaScript asynchronous. These technologies promote the construction of cloud and fog computing applications
leveraged from both consumers and creators for the material contribution.
Users of the recent product versions should not be installed but can be used
for software applications by The Web 2.0., using lightweight modeling
techniques, incorporating and summarizing existing tools and services and
allowing simplified connection to the web, using dynamic and user-
friendly platforms.

- Service-oriented computing, which is a key paradigm of comparison in
both cloud and fog computing environments, facilitates device and app
creation by using "services" as the key element. The architecture of reduced interconnected, evocative software and services is based on service-
oriented computing. Site confidentiality (the exact position of a resource
cannot be communicated to the customer or consumer), close connection,
program language-independence and sustainability, are the core components of a service.

- Web Services Description (WSDL) and Simple Object Control Protocol
(SOAP) interaction frameworks enable various applications to access the
Internet infrastructure platforms through cloud computing and seriously
reconsider defining the documentation universal Internet services. SP.NET
and ADO.NET maintain interactive features through the Internet service
networks.

- Resources operated and supplied by/to several customers with a metered
infrastructure with various granularities provided a preferred cloud-
compliant SLA. In the EU the consortiums are part of the travel industry
and developmental literature devoted attention to developing open-source
platforms and technologies that are the key to developing cloud and fog
computing systems for the future [1].

10.8   CONTRADICTION IN CLOUD COMPUTING AND FOG COMPUTING

The several issues involved with the implementation of fog computing must be
considered to accomplish the aims of fog computing are as follows.

10.8.1   NETWORK OF FOG

The network, which is central to the connection of the various heterogeneous
modules (devices and devices) to the specified services, is the complex fog

computing implementation that is one of the main components. New techniques, such as SDN and Channel Function Virtualization, can be used to accomplish infrastructure efficiency and cost savings. Defined Network Software (NFV). The fog register allows the connected interfaces to pick the optimal wireless connection speed to facilitate traffic privilege reservations, network virtualization, etc. The NFV's main feature is to substitute software applications for the network functions. The efficiency of virtualized network devices is a crucial issue that needs to be addressed urgently [4].

10.8.2   QUALITY SERVICE (QoS)

Network infrastructure efficiency of the NVS is a very significant indicator of QoS. It is determined by connection, power, quality and latency. The four metrics of QoS are measured and recorded [4].

10.8.3   INTERFACE AND PROGRAMMING MODEL

As services are obtained via virtual networks within a fog computing network, robust and dynamic interface and planning models are required. Current thinking indicates a better future for app-centric computing in fog computing to create possibilities for application-aware modules in the fog computing context that allow for many forms of optimization. Successfully, functional frameworks that are available on many heterogeneous platforms are difficult to construct [17] and have established higher-level programming models for new web apps; more general systems need to be established.

10.8.4   COMPUTER OFFLOAD

Computer offload helps to preserve computing efficiency. Extended battery capacity or supported memory sharing on multicore applications are important.

10.8.5   DELIVERY AND RESOURCE MANAGEMENT

The implementation delivery is disrupted as the end nodes in fog's computing system switch because of a dynamic metric, such as throughput, memory and bandwidth. In mobile crowd-sourcing or sensing apps, fog computing and IoT have a new critical position. In fog computing settings, the exploration of resources in extremely complex knowledge resources and the exchange thereof resources is of fundamental importance. Information sharing is a central feature to be recognized in computational and storage environments to meet the various challenges of fog computing [4].

10.8.6   CONFIDENTIALITY AND SECURITY

While many researchers have devoted great efforts to cloud storage safety concerns, there are few fog computing initiatives and efforts. Suspicious malicious activity, denial of service (DoS) and packet sniffing, for example, can be detected in customized applications on geodispersed fog computing systems. Protection problems in the hardware and software layers must be investigated.

The development of the required degree of fog computing interoperability support access tools is another important task for fog computing. Given the multidimensional nature of fog computing's multiple bodies, this involves a collective solution. The need for an open architecture that would not rely on the technological advance and the kind that can be made in the fog computing world is one approach to achieve compatibility. A single common interface will include fog applications, will improve the accessibility and use of fog computing and will potentially increase fog computing efficiency and innovation [21], as a part of an open architecture that will greatly reduce the costs of developing fog applications. Multitenancy, which already exists for heterogeneous agents that control infrastructure in a fog computing environment, is one of the main problems of fog computing, differentiated by differing degrees of safety, services, regulation and legislation. The problem of protection cannot be overemphasized in a multi-tenant setting. One of the keys needs to be identified and access control to implement the required degree of protection (IAM). IAM includes an interpretation of the IAM as part of business items in a machine, e.g., a computer, the SIM card, ISO/IEC 24760-1. The identity indicator, which is successful and connected to an object, is a reflection of the data used to authenticate an entity, such as login, lock, password, smartcard. The identification is an individual that identifies an individual entity. By constantly addressing reliability and effective questions, the challenges of fog computing can be exacerbated. Solid identity control is the central hypothesis in achieving complex safety management in noisy computer operating systems.

10.9 LEGAL DIMENSIONS OF CLOUD COMPUTING AND FOG COMPUTING

As regards the proliferation of service providers in any country, law and order are essential to ensure that cloud and fog processing technologies are healthy, including a multi-renter concept is implemented. Certain administrative metrics are as follows in cloud and fog computation situations:

1. Cloud computing availability is becoming more competitive. Promoting some cloud providers aim at providing services that result in locking, blocking and reliance on consumers. This is a challenge caused by the different data types and implementations that are accessible to cloud systems and infrastructure. Given the world in which cloud computing exists, it takes regulatory mechanisms to support consumers of cloud technology. In addition, regulatory provisions can enhance the delivery of cloud applications on flexible interfaces such that customer lock-in problems have been left behind [18] in the past.
2. Legal provision is necessary if either a customer or a supplier wishes, without fines, lack of data or any other unintended effect, to partner with the firm [18].

3. Must be visually stunning to ensure that all cloud and fog service providers have a certain level of consistency enshrined in service-level agreements. This will remove much of the inadequate or untrustworthy data service management already provided by many of the fog or cloud services.
4. To ensure that the information that they carry within virtual server storage confines is not being exploited at any cost to disadvantage the data user, the service providers must be supervised.
5. Other logistical concerns in the architecture of cloud computing need to be examined, including legal setups such as the geographic location of information, legal standing in the field of the cloud service provider and whether the cloud provider may use a non-country-based architecture; what happens with the data after the deal expires, etc.

It is necessary to remember that all the aforementioned points depend upon the need to allow cloud storage to be interoperable and reversible. Many cloud computing providers of all kinds of platforms need the interoperability of cloud service. In all circumstances, it is appropriate to promote the adoption of the policy cooperation of the legal and technological aspects of cloud computing.

10.10 CONCLUSION

In our everyday lives, fog computing systems have become significant. From the above conversations, it is clear that the potential of fog computing will be ubiquitous and prevalent in all fields, social and business. Also, earthen objects are possible through the use of simple computer equipment and integrated computing capabilities, to achieve high-caliber data computationally. The beauty of this innovation is that this computer technology is now used on all intelligent devices, particularly mobile phones. Almost all of these mobiles also have browser apps completely installed. These include Google, music, song, movies, camera, schedule, alarms, Facebook, Twitter, browsing, voicing, hang-outs, voicemail, WhatsApp, messages, e-mailing and so on. These apps include free-format videos with virtual viewing synthesis and associated multi-vision benefits that can be helped by edge technology. All of these applications include storage space, high-speed transmission and high-speed bandwidth linked in real time. Smart devices provide computing and residential monitoring as well as IoT environments via environmental intelligence (AmI) and are often used in homes, including in the case of home automation. In real-time cases, misuse of databases. All IT-related businesses develop IoT devices and services for computing, applications and networking in an interconnected fog. However, the basic foundations of fogging and its associated implementations have yet to be established by researchers to create traditional fogging models and architectures.

REFERENCES

[1] Madakam S, Date H (2016) Security mechanisms for connectivity of smart devices in the Internet of Things. In: Mahmood Z (ed), *Connectivity Frameworks for Smart Devices*. Springer, Cham, pp 23–41.

[2] Tseng YH, Lin CJ, Lin YI (2007) Text mining techniques for patent analysis. *Inf Process Manage* 43(5), 1216–1247.

[3] Jain A, Singhal P (2016) Fog computing: driving force behind the emergence of edge computing. In: *System modeling and advancement in research trends (SMART)*, international conference on IEEE, pp 294–297.

[4] Dang TD, Hoang D (2017) A data protection model for fog computing. In: *Fog and mobile edge computing (FMEC), 2017 second international conference on IEEE*. IEEE, pp 32–38.

[5] Bierzynski K, Escobar A, Eberl M (2017) Cloud, fog, and edge: cooperation for the future? In: *Fog and mobile edge computing (FMEC), 2017 second international conference on IEEE*. IEEE, pp 62–67.

[6] Bonomi F, Milito R, Zhu J, Addepalli S (2012) Fog computing and its role on the Internet of Things. In: *Proceedings of the first edition of the MCC workshop on mobile cloud computing*. ACM, pp 13–16.

[7] Vaquero LM, Rodero-Merino L (2014) Finding your way in the fog: towards a comprehensive definition of fog computing. *ACM SIGCOMM Comput Commun Rev* 44(5), 27–32.

[8] Madsen H, Albeanu G, Burtschy B, Popentiu-Vladicescu FL (2013) Reliability in the utility computing era: towards reliable fog computing. In: *Systems, signals and image processing (IWSSIP), 2013 20th international conference on IEEE*. IEEE, pp 43–46.

[9] Waheetha R, Sowmya F (2016) Fog computing and its applications. *Int J Adv Res Basic EngSci Technol (IJARBEST)* 2(19), October 2016.

[10] Zhu J, Chan DS, Prabhu MS, Natarajan P, Hu H, Bonomi F (2013) Improving web sites performance using edge servers in fog computing architecture. In: *service oriented system engineering (SOSE), 2013 7th international symposium on IEEE*. IEEE, pp 320–323.

[11] Ha K, Chen Z, Hu W, Richter W, Pillai P, Satyanarayanan M (2014) Towards wearable cognitive assistance. In: *Mobisys*. ACM.

[12] Zao JK, Gan TT, You CK, Chung CE, Wang YT, Rodríguez Méndez SJ, et al (2014) Pervasive brain monitoring and data sharing based on multi-tier distributed computing and linked data technology. *Front Human Neurosci* 8, 370.

[13] Yi S, Qin Z, Li Q (2015) Security and privacy issues of fog computing: a survey. In: *International conference on wireless algorithms, systems, and applications*. Springer, Cham, pp 685–695.

[14] Zhu J, Chan DS, Prabhu MS, Natarajan P, Hu H, Bonomi F (2013) Improving web sites performance using edge servers in fog computing architecture. In: *service oriented system engineering (SOSE), 2013 7th International Symposium on IEEE*. IEEE, pp 320–323.

[15] Mind Commerce (2017) Computing at the edge series: fog computing and data management. http://www.mindcommerce.com/files/FogComputingDataManagement.pdf

[16] Gupta H, Vahid Dastjerdi A, Ghosh SK, Buyya R (2017) iFogSim: a toolkit for modeling and simulation of resource management techniques in the internet of things, edge and fog computing environments. *Softw Pract Exper* 47(9). pp 1275–1296.

[17] Madakam S, Ramaswamy R (2015) 100 new smart cities (India's smart vision). In Information technology: Towards new smart world (NSITNSW), In: 2015 5th National Symposium on IEEE. IEEE, pp 1–6.

[18] Gubbi J, Buyya R, Marusic S, Palaniswami M (2013) Internet of things (IoT): a vision, architectural elements, and future directions. *Future Gener Comput Syst* 29(7), pp.1645–1660.

[19] Khan R, Khan SU, Zaheer R, Khan S (2012) Future internet: the internet of things architecture, possible applications and key challenges. In Frontiers of information technology (FIT). In: *2012 10th international conference on IEEE*. IEEE, pp 257–260.

[20] Khattab J (2017) Fog computing. https://www.slideshare.net/joudkhattab/fog-computing78498759

[21] Bilal K, Erbad A (2017) Edge computing for interactive media and video streaming. In: *Fog and mobile edge computing (FMEC), 2017 second international conference on IEEE*. IEEE, pp 68–73.

[22] Abdelshkour M (2015) IoT, from cloud to fog computing. Cisco Blog-Perspect. https://blogs.cisco.com/perspectives/iot-from-cloud-to-fog-computing

11 Role of Virtual Reality and Multimedia Computing in Industrial Automation System

Umesh Kumar Lilhore[1], Sarita Simaiya[1], Leeladhar Chourasia[2], Naresh Kumar Trivedi[1] and Abhineet Anand[1]

[1]Chitkara University Institute of Engineering and Technology, Chitkara University, Punjab, India
[2]SAGE University, Institute of Computer Application, Indore, India

11.1 INTRODUCTION

Production industries are currently attempting to increase global productivity by merging industry with digital technologies. Virtual reality (VR) is being employed in industrial companies' production procedures as a useful tool for achieving quick knowledge aggregation and policy decision designing using imagery and practical experience [1]. The U.S. Ministry of Defense's Virtually Production Strategy helped bring virtual production technology and automation into establishment in the early- to mid-2000s. Manufacturing companies can use virtual reality to model the mass production procedure and manufacturing process setups, allowing them to spot potentially harmful scenarios. VR may also be utilized to involve an individual in a projected workplace, with the motion captured to assess task capability and competence.

Multimedia computing involves media, graphics, sound and three-dimensional views of an object play a vital role in VR. The VR renders any object into a computerized digital system based on three dimensions. One may simply demonstrate an object and also its elements, allowing users just to see how they operate. Technology accelerates the operation of engineers and technicians creating virtual design/models of modern vehicles, machinery and robotics. However, it isn't all. Manufacturers of manufacturing processes, who are progressively adopting virtual exhibitions, adopt this approach [2]. VR technology, on the other hand, is better than just the one can still accurately reproduce the company culture and sometimes even environmental conditions. Everything is done to offer the greatest possible circumstances for staff development and also to instruct workers about how to move.

DOI: 10.1201/9781003196686-11

Production involves the procedure of converting raw components into finished goods using manpower and industrial equipment. Fresh produce, machinery, clothing, all polished metals, timber goods, etc. are examples of products. Such items may be shipped to some other producer to be used in the creation of more complicated items, and they might be sold. Environmental and physical methods can also be used in manufacturing [3]. VR's capacity to react to a range of outcomes even outside the surroundings provides one a benefit under certain tactical situations. Additional uses that contribute to the advantages of just using VR include designing and evaluating items that you wouldn't even be able to view anything unless attending the designer's premises. One can communicate with goods using this technique despite having to participate with people.

This chapter represents the importance of multimedia commuting, virtual as well as augmented reality techniques, different aspects and practical limitations for actively supporting the developers for actually creating industrial environments that generates accurate and feasible experiments of the behavior and attitude of machinery and real procedures and reliable performance of remotely believable methods [4]. The complete chapter is divided into various sub-sections that include survey work, virtual reality, mobile computing overview, applications in industrial automation and challenges and conclusions.

11.2   LITERATURE SURVEY

Multimedia computing with virtual reality has been used as a powerful method to understand a concept quickly in numerous disciplines. However, if this data is given in diverse ways, individuals are now more adaptable to specific knowledge and build simpler computational methods.

The following keywords are suggested by various researchers in the field of Industrial automation using multimedia computing and virtual reality.

Ahmed et al. [1] worked in the field of VR in industrial automation. The discipline researchers now recognize VR as technology, an interdisciplinary research branch of computation that arose in the 1980s from the development of 3D interactive graphics and automotive simulations. Virtual reality can assist automakers in reducing cost expenses associated with traffic manufacturing and installation. When used for teaching, interactive VR features allow automakers to significantly accelerate and enhance the classification task, enhancing efficiency. It's the most important aspect of VR and AR application development; it also can greatly impact the experiences. The system can be accessed well in a virtual environment with enhanced reliability and function faster at executing tasks by combining AR-VR with engagement and gasified architecture. It's also worth noting that gaming for such a product is not the same as loyalty marketing. Relatively brief concentration can be aided by intrinsic motivations and reinforcement schemes. Intrinsic rewards are extremely important when it comes to developing the best experience for a casual relationship. A well-designed environment can make even the most mundane activities entertaining.

Herwan et al. [5] worked on smart manufacturing. Implementation of VR enables the developers' administrators to replicate lean manufacturing layouts and processing equipment, allowing them to spot various dangerous scenarios. Numerous industrial information technology is deemed quest, meaning that their failure might result in devastating repercussions in terms of morbidity and mortality or property damage. As a result, extreme caution must be used when designing them to ensure that they are perfect. Despite this, complex processes are frequently used to ensure that any un-anticipated events may be dealt with predictably. Emergency failure tolerance due to equipment/software failures must frequently be factored in.

Liu, et al. [6] worked on multimedia in virtual reality. The maintenance team and manufacturers may use virtual reality to go towards any IoT-enabled device and recognize an item to obtain any knowledge from either a corporation's back-end organizational structure. VR technology has a significant impact on the way individuals perform, interact and offer services. Indeed, technology has the potential to dramatically alter – and extend – whatever humans are conscious of. The effects are expected to be significant in the case of robotics. It can take place in many circumstances. Optimization of manufacturing applications is a prerequisite for the IR 4.0 goal to become a reality. There have been a number of reasons that can enhance the odds of automation initiatives succeeding. A survey of the literature on justification strategies and their use is conducted. Another collection of tools for determining explanation approaches is established based on the review. The model is tested in a firm that is working on a manufactured process improvement plan that requires selections on processes and systems that must be evaluated and justified in a methodical manner. Throughout the proposed development, several rationalization approaches were chosen and evaluated.

Majumdar, et al. [7] worked on industrial automation and the role of virtual reality. Information systems provide combined access to a variety of data sources by simulating different emotions using technology. VR technology information systems can blend direct struggles with machine content to provide virtually lived emotions. Whenever it concerns educating people to handle machines and execute procedures, virtual reality offers up a world of opportunities. In instructions and demonstrations, always so far which can be explained. VR provides many of the rewards of realism even without expenses or risks connected with only a journey to the factory to attain a certain purpose. Digital transformation is unique to each operation. For example, throughout the corn chips production industry, robotics are used to cook chips that have already been chopped and cleaned entering the processing chambers. Smart manufacturing, on either side, is the total automation of a corporation's entire process with no human intervention, with the only function a person does being to control and maintain the entire mechanization.

Pratticò, et al. [8] suggested the application of industrial automation. Production is among the most significant potential applications for automation technologies. Several people associate robotics with automation systems. Machine transmission systems in the automobile sector, automated production machinery and some natural processes are typical application automations. While just VR technology is still very much in infancy, its great promise is being felt in a variety of sectors throughout the globe. This will ultimately help learners from everywhere on the globe to communicate and collaborate as if they're in the same area, engaging in live time between everyone on 3D and employing materials and equipment that just don't exist.

Kovalenko, et al. [9] suggested virtual reality in industrial automation. The term "automation" is formed from the Grecian terms "auto" (individual) and "matos" (device) (moving). As a result, automation seems to be the method for technologies that "move on their own." Aside from the fundamental interpretation of the phrase automated systems attain considerably significantly higher levels of power, accuracy and fast response over manual methods. Information systems are widely used in industrial automation. The inclusion of automating and robotics supervision in corporations and industries like leather or paper sectors is referred to as business automation. Automation is, from the other hand, primarily softer manufacturing, which includes back operations finance, etc. Although both industry and processes management need automation, its major applications are distinct.

Morris, et al. [10] suggested the multimedia importance in smart manufacturing. Industrial automation processes become more advanced following best practices and methodologies they employ, as they expand to include bigger functions of the organization, like different components or the whole production facility, and even multiple factories. But as they incorporate industrial production with certain other lines of employment, including dealing with customers, funding and the global supply chain including its company, the use of these will increase. Comparatively low-cost robots, on the other hand, only interact with a single computer or, at maximum, a cluster of equipment, rely on technology, circuits and integrated computers rather than IT.

Raman, et al. [11,12] suggested the challenges of a manual manufacturing system and its solution. Production and design standards, working conditions and price competition, production costs and jobs all influence the level of technology required for a given production facility. It's worth remembering that the cost of automating must be compensated by the gain in profitability. Businesses may integrate classrooms as well as on training inside a simulated environment, wherein users may educate about products/machinery and then practice on interactive reality machines. Further repeats can be performed, that is an excellent practice that helps to develop muscle ability of the brain to be doing tasks quickly.

11.3 VIRTUAL REALITY AND MULTIMEDIA COMPUTING

As the title implies, "industrial automation" refers to the use of advanced manufacturing computing to automated manufacturing operations. Industrial computing can securely manage automation, assembling, inspection and online monitoring for all other applications due to the availability of the latest computer technologies.

Virtual reality (VR) provides the computer-generated environment that appears to be reality. In production, scheduling and the decisions to reorganize equipment in the region can all be carried out completely before employees move equipment [13]. VR technology may allow companies to improve their brand almost as much as augmented reality. In contrast to the enlarged reality that provides a real-world perspective, VR is also a fully computerized atmosphere. It becomes possible to change everyone from the environment to an object whereby the employee engages mostly in the program. In particular, for planning and training scenarios, VR gives more versatility [14].

"Multimedia computing (MC) technology is a modern innovation that's already opening up an entirely new of computer possibilities. A fusion of video, music, 2-D, 3-D animation, and image with graphics, text and advanced computers is what MC is all about" [15].

11.3.1 How Does VR Technology Work?

VR technology is made up of hardware and software techniques that allow people to communicate to engage someone in a virtualized world. The key idea behind VR is to build a user-virtual reality relation. Eventually, virtual things are sensed by human sensory technology [16].

11.3.2 Key Principle of VR

To engage the viewer inside a virtual world, VR interacts alongside them. The idea is to develop communication among the source and even a virtual world. This necessitates the use of program programs such as visual graphics, some real-time computations and hardware applications such as computer interfaces. Using improved sensory methods, the results of communication between all the characters and also the virtual world are returned to a recipient's perceptions. That engagement cycle must be carried out with latency equal to an aspect ratio of the aroused relevant sense called the "real-time" process [5,17].

Finally, by utilizing techniques that impact physical perception, virtual environments are experienced. Visual tools, auditory response, mechanical and pressure responses and movement platforms seem to be examples of such innovations. Communication in a virtual space is achieved by a link among the entire virtual world and also the participant's shared network. The idea of sensation of appearance is used to assess a customer's experience in such a virtual world. Table 11.1 shows the relationship between physical sensations of the body and VR advancements. This is based on how well the client may tour the virtual world, various options it has towards interaction and also the sensory input offered by the technology. Natural participation in the digital immersing system is crucial to the realistic experience [18].

11.3.3 Issues with VR Technology

The individual is involved within the procedure because their perspective seems to be the consequence of alchemy among their involvement, various techniques and also the project. While building a VR program, the entire structure comprised of humans and robots must be addressed. As a result, social errors, technology and also the app are important considerations in virtual immersion. Several relevant elements include technological and scientific difficulties that are collaboratively handled by a research world concerned with technologies such as computer animation and microelectronics, and even a research group dealing with human aspects (ecology, neurology, including social psychology) [19]. These four primary characteristics might impact the

TABLE 11.1

A Relationship among Physical Sensations of Body and VR Advancements

Key Factors of Human Sensation	Features	VR Advancements
Sense of vision	Between 450 and 780 nm, it's responsive. Brightness ranges between 0.02 cd/m^2 to 1 million cd/m^2. High-definition. Sensitive to contrasting colors VR requires 64 Hz each vision 180°c Vision in the horizontal direction, a rotating rate equivalent to 800°/s Inside the vertical position, the FOV is 140 degrees.	Interactive display. Properties for sharpness. Eyes tracing on a huge monitor with a wide field of view. The screen that is worn on the face.
Stereoscopic vision	Region for the simultaneous projection (Panum zone) Connection between visual adaptation and converging	Splitter method for right and left streams. The mechanism for monitoring the range of motion.
Sound and waves	A frequency band of 1–30 Mhz with high definition Modeling in both ears.	Monitoring of the ears. Outer ear transmission feature of the user. Sequencing in actual environments. Visualization at high frequencies.
Ears	Within space, there is an accelerating meter mainly delays with translation and rotation. In the translation process, the sensory limit is 5 cm/s^2. In the rotation process, the sensory limit is 2°/s^2. Exceptional sensitivities.	Platforms that are dynamic.
Perceptible by touch/ tactile	Scalp sensors with a sensitivity of 100 kHz.	Sensory inputs such as a vibration sensor, electricity.

participant's sense of existence: (a), system responsiveness in response to the recipient's mobility (b), multisensory feedback loops and also the strength of the sensual connection (c), subject involvement within virtual space (d) and environmental connections.

11.3.4   RELATIONSHIP AMONG MULTIMEDIA COMPUTING AND VR TECHNOLOGY

Multimedia shows its true ability to inspire consumers and capture their interest; both of these constitute key traits while attempting to give a better level of interactivity and developing sustainability with Virtual World programs. VR technology is just an extension of a multimedia technique that employs fundamental multimedia components including visuals, audio and motion. As it needs navigational input from either a human, VR technology may also be considered an immersive audiovisual in its most comprehensive form. The VR technology mainly includes the five multimedia components: text, picture, video, audio, motion and animation [20].

11.4   KEY TECHNOLOGIES OF VR

In both the physical and virtual worlds, communication between such a person and a virtual world involves a range of devices. Computer programming, living thing interactions and informatics are responsible for a lot of these innovations [6].

11.4.1   SENSOR INTERFACE AND MOTION TRACKING

Motion capture devices are also employed to animate the recipient's avatar simultaneously and also in cooperation with the recipient's movements. Such techniques try to notify the computers about the participant's movements. The idea is to connect a digital environment connection to something like a modern world connection. Dynamical, electromagnetic, ultrasonic and image-based methods are all available. VR devices almost always employ structured techniques rather than image recognition. The eye locations are calculated based on hand placement, allowing yet another virtual lens to be placed for every vision to generate the virtual picture based on the recipient's location in the virtual space [21].

11.4.2   VISUALIZATION COMPONENTS AND DEVICES

To create a picture, visualizing systems employ a variety of techniques. These may display images and videos or utilize an electronic touch screen. The first form of visualization process is defined by a fixed panel beside whereby the customer stands. Monitors come in a variety of widths and lengths. Enormous displays are important for rendering virtual environments at one dimension for viewers so that they can then view, traverse and communicate with them even though they would be in actual life [22,23].

11.4.3   3D SOUNDS

The utilization of 3D sound instead of in VR technique mainly improves the recipient's absorption by providing value while dealing with virtual items and adding to the aural atmosphere in the individual user scene. To enable the consumer to determine the type of audio occurrence in the virtual world, any three-dimensional sounds must be exact. In the virtual world, 3D audio technology consists of a genuine head platform, an earphone or even a multi-canal device with a software-based, three-dimensional audio computation structure. Image-rendering philosophies are used in source code devoted to the actual computation of 3D audio. Audio analogies can be intended as an economic benefit for sounds to enhance the accuracy of audio specificity.

11.4.4 GRAPHICS AND ANIMATION

Usually, numerous activities are flexible in today's environment. With time, things develop. This is much more likable to simulate these occurrences and activities is a three-dimensional authentic computerized animation. This will be the primary medium; however, this will offer a new aspect to multimedia throughout the coming years, along with many other mediums like artificial audio, voice and music. In addition, this multimedia strategy also requires participatory tools. The introduction of strong 3D gadgets especially provided a fresh perspective on the virtual world. Now even the animation can enter, appreciate, change and fully experience the artificial environment an individual has built [7].

11.5 CHALLENGES IN MANUAL INDUSTRIAL SYSTEM

Although the industrial sector is vital, it also has several issues.

11.5.1 TRAINING FOR THE MANUFACTURING SECTOR

Each worker must get appropriate production training. When they do possess appropriate experience and competencies how soon are they capable of handling all types of machinery and making things? However, providing industrial sector training to every individual is quite difficult. It can be pretty hectic at moments. Since many organizations aren't limited to a single place, maybe they just have multiple operation halls, various towns and even other nations, manufacturing firms must provide separate training programs at various locations. As a result, businesses must spend more time and resources on training courses [24].

11.5.2 WORKPLACE ACCIDENTS AND INJURIES

Industrial production has a higher rate of accidents than other industries [25]. Workplace injuries can happen for a variety of causes. The following are by far the most prevalent factors:

- Devices may fall, particularly inside the factory.
- Breakdown of machinery is by far the most common reason for incidents. Equipment malfunction can strike at any point, posing a risk to workers' lives.
- Extreme amounts of vibration in machines, which can cause deafness.
- Exposure to hazardous substances can lead to a wide variety of mishaps and life-threatening situations.

11.5.3 LATEST TECHNIQUES/INNOVATIONS MUST BE ESTABLISHED IN THE INDUSTRY

Understanding and innovative things practice is often tough. An individual has to go through the enlistment process and gain new skills. Organizations also require talented and qualified staff. Choosing the correct ones, on the other hand, becomes quite challenging. This is a lengthy procedure. Companies' primary goals are to save costs and resources. Organizations aim to provide items as rapidly as feasible without raising their obligations, while still maintaining the health and protection of workers [26].

There are several options to accomplish the above challenges [27]:

- **Use precise and effective instruments:** The very first step is to employ precise and effective instruments that may assist businesses in meeting market demands.
- **Staff can be trained using e-learning approaches:** Staff can be trained in the same way via a digital platform. It is not necessary to collect all of the personnel in one location. This will reduce both effort and resources and it'll be very effective during the pandemic season.
- **Over periodic intervals, inspect the device:** Operate a factory that follows all security regulations and laws. Renew the expertise of your personnel. Injuries will be less likely as a result of this.
- **In the industrial industry, VR technology can be useful:** One may even be startled to learn that VR technology is the industrial sector's innovation. Several firms produce use VR.

11.6 INDUSTRIAL AUTOMATION SYSTEM

Industrial automation (IA) is an integration of operational schemes that includes high computing processors, robotics and software innovations to assist humans in different mechanisms and machines in an organization. It will be the next phase in the industrialization process after mechanization [8].

11.6.1 BENEFITS OF INDUSTRIAL AUTOMATION

An industrial automation (IA) provides the following benefits [28]:

- **Reduces operational expenses:** An IA reduces capital investment, as well as paid sick leave and vacations, which would otherwise be incurred by human intervention. Furthermore, smart manufacturing does not necessitate better employee perks including incentives, retirement cover and so on. Most of all, considering the large upfront investment, it decreases the employees' yearly salary, resulting in a significant saving for the firm. Routine maintenance costs are lower for equipment utilized in process control since it fails less frequently. Both computers and service experts are needed to fix it if it crashes.
- **Productivity is significant:** Even though many firms recruit dozens of staff members for two to three schedules to operate the factory for both the optimum amount of hours, its facility might still be closed indefinitely. Smart manufacturing helps the firm achieve its goal by enabling it to operate a

manufacturing plant 24/7 a day, 7 days a week, even 365 days annually. It results in a considerable increase in the company's performance.

- **Improve quality:** The error associated with a human being is reduced by high-quality automation. Furthermore, unlike humans, robots do not experience tiredness, resulting in items of uniform quality created at different times.
- **Great flexibility:** Introducing a new activity to a production line necessitates learning with a single user; conversely, robotics may be taught to perform any operation. It increases the production program's flexibility.
- **The high degree of quality of information:** By using automation, businesses can acquire critical brand information, optimize resource quality and lower data gathering expenses. It gives us the information needed to make informed choices about removing defects and optimizing our procedures.
- **High safety:** By employing robotics to tackle hazardous circumstances, smart manufacturing may increase the production process safer for humans.

11.6.2 KEY CHALLENGES IN INDUSTRIAL AUTOMATION

The following possible challenge can occur [29]:

- **Initial phase investment is expensive:** An early expenditure required to transition from just a manual manufacturing plant to an automated manufacturing plant is significant. In addition, there are huge costs that include adequate training and development staff to use this more advanced technology.

11.7 VR APPLICATIONS IN INDUSTRIAL AUTOMATION SYSTEM

While more and more computerized automation application scenarios emerge, intelligent production and, as just a result, through the use of computing in production industries, has seen a constant increase. As per a BDO analysis, one out of every four industries was using Industry 4.0 approaches at the beginning of 2020, a considerable increase from only sixmonths before, when the percentage was closer than 7%. Such facts are not unexpected, however, given how computing's increasingly important function in production can be seen at almost every stage of industrial manufacturing [30].

11.7.1 SMART MANUFACTURING

Manufacturing computer systems are recommended for application production that goes beyond human strength. Within industries, commercial PCs feature cutting-edge setups and quite durable modules of the system are utilized to precisely produce and install complicated technical elements, as well as execute automatic quality assurance at fast rates. The emergence of durable and efficient business processes has benefitted the automotive and specialist equipment manufacturing sectors significantly [31].

11.7.2 Asset Management and Tracking

Among the most important advantages of manufacturing computing is its capacity to analyze any particular task or equipment for potential anomalies. Industrial computer systems being widely utilized in all daily operations of businesses including retail and medical, and that's not restricted to quality inspection. This is attributable to the fact that devices make performance monitoring easier, and most virtually, and the precision and reliability with what they accomplish renders them indispensable [32].

11.7.3 Visualization and Simulation

Today's technological discoveries owe a large part of their effectiveness to the business process. Equipment like "din-rail computers," when used in combination with edge computing technology, may form a highly sophisticated automation solution that will allow instruments and industrial operations to be rigorously evaluated in simulating real-world scenarios. It enables risk analysis, extrapolation of potential and testing of every technology inside any environment with next to no danger of negative consequences or capital risk [10].

11.7.4 Workers' Safety

Industrial computer systems can be used to improve industrial security in the worksite in the frequent production environment. A fanless variant minimizes trash, humidity and pathogens from spreading to basement workers. It's not just that; production managers need to set guidelines for managing personnel to sanitize and wash equipment once it has been often used by using a touch device on the Internet too. Such alerts can indeed be configured to go off when IoT senses that a device is being used, or perhaps once it has been utilized [11].

11.7.5 Automation of Documentation Work

Nevertheless, several ongoing advances in production, documents and necessary documentation are rather important for the profession. But, as manufacturers become self-sufficient, the sector has advanced to the point where many are already embracing the commercial industry of research and development too. Towards that objective, industrial computers are used in combination with automation to simplify procedures like BOM preparation and invoice processing [9].

11.8 CONCLUSIONS

Industrial automation is gaining popularity in a wide range of sectors due to the numerous advantages it provides, including higher efficiency, reliability and protection at an affordable rate. The VR technology of individual operations (for example, development, manufacturing, medical treatment, research and air strikes) frequently focuses on such an automatic interaction among VR technology and the theory and practice of these tasks. This chapter primarily focuses on the significance of virtual reality technologies and the development of this platform. While this scope and diversity of methods is vast, two illustrative domains (manufacturing implementations and industrial uses) are thoroughly discussed to provide an understanding of the scale of the advantages and the obstacles that remain.

This chapter represents the importance of multimedia commuting, virtual reality techniques and their different aspects and practical limitations for creating industrial automation environments. In this chapter, we analyze difficulties and explanations on how the combination of graphics, virtual reality and vision as well as communication innovations can improve the importance of carriers in this modern context. Another way VR technology is influencing production is through increasing safety. Manufacturing companies may utilize VR technology to simulate automated assembly layouts and manufacturing techniques, allowing managers to spot extremely hazardous conditions. Industrial automation provides numerous benefits in the manufacturing sector.

REFERENCES

[1] Ahmed, Salman, Lukman Irshad, and H. Onan Demirel. 'Prototyping Human-Centered Products in the Age of Industry 4.0'. *Journal of Mechanical Design* 143, no. 7 (2021): 1–15.

[2] Al-Turjman, Fadi, and Sinem Alturjman. '5G/IoT-Enabled UAVs for Multimedia Delivery in Industry-Oriented Applications'. *Multimedia Tools and Applications* 79, no. 13–14 (2020): 8627–8648.

[3] Berg, Leif P., and Judy M. Vance. 'Industry Use of Virtual Reality in Product Design and Manufacturing: A Survey'. *Virtual Reality* 21, no. 1 (2017): 1–17.

[4] Cesar, Eduardo L., Gustavo S. Fernandes, Marcelo T. N. Kagami, and Talles N. Calisto. 'Technological Obsolescence Management: Monitoring Electrical Equipment and Automation Systems'. *IEEE Industry Applications Magazine* 26, no. 4 (2020): 82–87.

[5] Herwan, Jonny, 'Comparing Vibration Sensor Positions in CNC Turning for a Feasible Application in Smart Manufacturing System'. *International Journal of Automation Technology* 12, no. 3 (2018): 282–289.

[6] Liu, Bingchun, Mingzhao Lai, Jheng-Long Wu, Chuanchuan Fu, and Arihant Binaykia. 'Patent Analysis and Classification Prediction of Biomedicine Industry: SOM-KPCA-SVM Model'. *Multimedia Tools and Applications* 79, no. 15–16 (2020): 10177–10197.

[7] Majumdar, Abhijit, Himanshu Garg, and Rohan Jain. 'Managing the Barriers of Industry 4.0 Adoption and Implementation in Textile and Clothing Industry: Interpretive Structural Model and Triple Helix Framework'. *Computers in Industry* 125, no. 103372 (2021): 103372.

[8] Prattico, F. Gabriele, and Fabrizio Lamberti. 'Towards the Adoption of Virtual Reality Training Systems for the Self-Tuition of Industrial Robot Operators: A Case Study at KUKA'. *Computers in Industry* 129, no. 103446 (2021): 103446.

[9] Kovalenko, Ilya, Efe C. Balta, Yassine Qamsane, Patricia D. Koman, Xiao Zhu, Yikai Lin, Dawn M. Tilbury, Z. Morley Mao, and Kira Barton. 'Developing the Workforce for Next-Generation Smart Manufacturing Systems: A Multidisciplinary Research Team Approach'. *Smart and Sustainable Manufacturing Systems* 5, no. 2 (2021): 20200009.

[10] Morris, K. C., Yan Lu, and Simon Frechette. 'Foundations of Information Governance for Smart Manufacturing'. *Smart and Sustainable Manufacturing Systems* 4, no. 2 (2020): 20190041.

[11] Raman, Arvind Shankar, Karl R. Haapala, Kamyar Raoufi, Barbara S. Linke, William Z. Bernstein, and K. C. Morris. 'Defining Near-Term to Long-Term Research Opportunities to Advance Metrics, Models, and Methods for Smart and Sustainable Manufacturing'. *Smart and Sustainable Manufacturing Systems* 4, no. 2 (2020): 20190047.

[12] Saleeby, Kyle, Thomas Feldhausen, Lonnie Love, and Thomas Kurfess. 'Rapid Retooling for Emergency Response with Hybrid Manufacturing'. *Smart and Sustainable Manufacturing Systems* 4, no. 3 (2020): 20200050.

[13] Dănuț-Sorin, Ionel R., Constantin Gheorghe Opran, and Giuseppe Lamanna. 'Lean 4.0 Dynamic Tools for Polymeric Products Manufacturing in Industry 4.0'. *Macromolecular Symposia* 396, no. 1 (2021): 2000316.

[14] Ghobakhloo, Morteza, Masood Fathi, Mohammad Iranmanesh, Parisa Maroufkhani, and Manuel E. Morales. 'Industry 4.0 Ten Years on A Bibliometric and Systematic Review of Concepts, Sustainability Value Drivers, and Success Determinants'. *Journal of Cleaner Production* 302, no. 127052 (2021): 127052.

[15] Gonçalves, A., M. F. Montoya, R. Llorens, and S. Bermúdez i Badia. 'A Virtual Reality Bus Ride as an Ecologically Valid Assessment of Balance: A Feasibility Study'. *Virtual Reality* (2021). 10.1007/s10055-021-00521-6.

[16] Guo, Ziyue, Dong Zhou, Jiayu Chen, Jie Geng, Chuan Lv, and Shengkui Zeng. 'Using Virtual Reality to Support the Product's Maintainability Design: Immersive Maintainability Verification and Evaluation System'. *Computers in Industry* 101 (2018): 41–50.

[17] Holzwarth, Valentin, Johannes Schneider, Joshua Handali, Joy Gisler, Christian Hirt, Andreas Kunz, and Jan vom Brocke. 'Towards Estimating Affective States in Virtual Reality Based on Behavioral Data'. *Virtual Reality*, (2021). 10.1007/s10055-021-00518-1.

[18] Ko, Ginam, placeB Inc, Guro-gu, Seoul, South Korea, Kyoo-Won Suh, and Industry Academic Cooperation Foundation, Hallym University, Chuncheon-Si, Gangwon-Do, South Korea. 'A Study on a Drawing Tool of a Spatial Drawing Application in Virtual Reality'. *International Journal of Multimedia and Ubiquitous Engineering* 13, no. 4 (2018): 7–12.

[19] Lee, Kyung-Hun, Ju-Yeong Kim, and Jae-Hyun Yoo. 'The Analysis of the Health Related Physical Fitness and Mental Health in Individuals with Intellectual Disabilities on Virtual Reality Exercise Program by Game Bike – a Pilot Study'. *Journal of the Korea Entertainment Industry Association* 14, no. 2 (2020): 119–129.

[20] Li, K., B. Yu, and Q. Li. 'Theoretical Analysis on High-End Equipment Manufacturing Growing: From the View of 3D-Spiral Technology Collaborative Innovation'. In *Multimedia, Communication and Computing Application*, 455–460. CRC Press (2015).

[21] Liu, Charles Z., and Manolya Kavakli. 'An Agent-Aware Computing Based Mixed Reality Information System'. *The International Journal of Virtual Reality: A Multimedia Publication for Professionals* 17, no. 3 (2017): 1–14.

[22] Loizeau, Quentin, Florence Danglade, Fakhreddine Ababsa, and Frédéric Merienne. 'Methodology for the Field Evaluation of the Impact of Augmented Reality Tools for Maintenance Workers in the Aeronautic Industry'. *Frontiers in Virtual Reality* 1 (2021). 10.3389/frvir.2020.603189.

[23] Lynn, Roby, Moneer Helu, Mukul Sati, Tommy Tucker, and Thomas Kurfess. 'The State of Integrated Computer-Aided Manufacturing/Computer Numerical Control: Prior Development and the Path toward a Smarter Computer Numerical Controller'. *Smart and Sustainable Manufacturing Systems* 4, no. 2 (2020): 20190046.

[24] Mountasser, Imadeddine, Brahim Ouhbi, Bouchra Frikh, and Ferdaous Hdioud. 'Big Data Research in the Tourism Industry: Requirements and Challenges'. *International Journal of Mobile Computing and Multimedia Communications* 11, no. 4 (2020): 26–41.

[25] Mubeen, Saad, Jukka Mäki-Turja, and Mikael Sjödin. 'Exploring Options for Modeling of Real-Time Network Communication in an Industrial Component Model for Distributed Embedded Systems'. In *Proceedings of the International Conference on Human-Centric Computing 2011 and Embedded and Multimedia Computing 2011*, 441–458. Dordrecht: Springer Netherlands (2011).

[26] Nassiri Pirbazari, Kameleh, and Kamran Jalilian. 'Designing an Optimal Customer Satisfaction Model in Automotive Industry'. *Journal of Control Automation and Electrical Systems* 31, no. 1 (2020): 31–39.

[27] Park, Ji-Su, Advanced Human Resource Development Project Group for Health Care in Aging Friendly Industry, Dongseo University, Dae-Kil Choi, Jae-Min Park, Yun-Seok Choi, Jae-Woo Suk, Hae-Mi Ji, and Tae-Hyung Yoon. '3-Sided Virtual Reality Mild Cognitive Impairment Rehabilitation Training Platform: An Update'. *Korean Journal of Neuromuscular Rehabilitation* 10, no. 1 (2020): 68–73.

[28] Schäffer, Eike, Maximilian Metzner, Daniel Pawlowskij, and Jörg Franke. 'Seven Levels of Detail to Structure Use Cases and Interaction Mechanism for the Development of Industrial Virtual Reality Applications within the Context of Planning and Configuration of Robot-Based Automation Solutions'. *Procedia CIRP* 96 (2021): 284–289.

[29] Shor, Daniel, Perceptual Intelligence Lab, Faculty of Industrial Design Engineering, Delft University of Technology 15 Industrial Design Landbergstraat, Delft, Zuid Holland 2628, Netherlands, Bryan Zaaijer, Laura Ahsmann, Max Wcctzel, Simon Immerzeel, Daniël Eikelenboom, Jess Hartcher-O'Brien, and Doris Aschenbrenner. 'Designing Haptics: Improving a Virtual Reality Glove with Respect to Realism, Performance, and Comfort'. *International Journal of Automation Technology* 13, no. 4 (2019): 453–463.

[30] Zhu, Wenmin, Xiumin Fan, and Yanxin Zhang. 'Applications and Research Trends of Digital Human Models in the Manufacturing Industry'. *Virtual Reality & Intelligent Hardware* 1, no. 6 (2019): 558–579.

[31] Simaiya, Sarita, Umesh Kumar Lilhore, Sanjeev Kumar Sharma, Kamali Gupta, and Vidhu Baggan. 'Blockchain: A New Technology to Enhance Data Security and Privacy in Internet of Things'. *Journal of Computational and Theoretical Nanoscience* 17, no. 6 (2020): 2552–2556.

[32] Lilhore, Umesh Kumar, Sarita Simaiya, Kalpna Guleria, and Devendra Prasad. 'An Efficient Load Balancing Method by Using Machine Learning-Based VM Distribution and Dynamic Resource Mapping'. *Journal of Computational and Theoretical Nanoscience* 17, no. 6 (2020): 2545–2551.

12 Virtual Reality–Based Education

Abhineet Anand[1], Naresh Kumar Tiwari[1], Umesh Lihore[1] and Rajeev Tiwari[2]
[1]Chitkara University Institute of Engineering and Technology, Chitkara University, Punjab, India
[2]School of Computer Science, University of Petroleum & Energy Studies, Dehradun, Uttarakhand, India

12.1 INTRODUCTION

It is crucial to understand why we first need to improve the quality of education before getting into the specifics of how VR in education might assist in enhancing the learning process. Historical facts and observations on the world have always been aimed in the majority of technology developed to facilitate learning. We had a wonderful tool before computers, which allowed us to keep facts: books. Books have transformed into e-books in the era of digital technology. Modern search engines make discovering information incredibly simple – you can obtain answers for many with only a few clicks [1].

The present approach to education has two key issues, although knowledge is easier for more people to use:

- Based on the same old framework – preservation of facts. Teaching approaches a focus on factual knowledge, yet access to and consumption of a lot of knowledge is not learning [2].
- Many individuals have difficulty understanding knowledge and being informed is not the same as being educated. There may be an overwhelming amount of information acquired in just a short time. They get bored, disconnected and frequently don't know why they initially learn about a topic [3].

In order to improve learning and engagement, virtual reality is employed. The method in which VR educational information is provided may be transformed; it works on the basis of constructing a virtual environment – real or thoughtful – that allows users not only to observe but also to engage with it. Taken into what you study, you are motivated to comprehend it entirely. To digest the information, it will need less cognitive stress [4,5].

The ability to provide a consistent and even training that is working for all your students or trainees is one of the most commonly stated issues for training managers

DOI: 10.1201/9781003196686-12

and departments. The question of different learning styles is one reason this might be so challenging [6,7].

Simply said, we don't all learn in the same manner, thus the best thing for a trainee can't work for someone else.

However, progress in technology of virtual reality opened the doors to a whole new manner of educating a variety of different groups of people – concurrently with different learning modes. There are four basic forms of learning – visual, auditional, reading and kinesthetics – powered by three of our main senses: sight, hearing and touch [8–10].

Virtual reality can help mixed groups to be educated since it gives all three sensory output kinds.

Indeed, the U.S. Department of Education, which is studying methods to employ the technology to aid learning disabled and high-functioning autistic students in schools nationally, has drawn attention to this all-encompassing sensory approach to learning using VR. In late 2018, the Office of Special Education and Programs (Office of Special Education and Programs) stated that it was investing $2.5 million in VOISS (Virtual Reality Oportunities to Implement Social Facilities) [11–13].

Yet VR has not yet been implemented into traditional classroom environments on a wide scale despite its potential as a training tool and its obvious enthusiasm with its possibilities of improving students' learning experience.

12.2 BENEFITS OF VIRTUAL REALITY IN EDUCATION

Virtual reality is an artificial computer world that you may experience and interact with. Every student understands with great fun and ease of complicated scientific concepts, such as when the student uses VR to understand the human body breathing system, and understands that so many gaseous elements enter the lungs as he/she respires and how oxygen is being filtered from the other elements and how they are binded to Red Blood Corpuscles (RBC). This oxygen-containing RBC penetrates the cell membrane to create water and carbon dioxide, thus breaking down glucose. The energy that is collected by the ATP molecule is released during this procedure. All of these explanations are available in 3D and 360°, helping the student to remember and recall effortlessly. The sample I've shown here is only a glance into the massive library of our VR material. Virtual reality is utilised in education to increase student involvement. The real world around us is created by virtual reality. The manner of education delivered currently may be transformed. It can also be utilised in industry, medical, safety and military, counterter-terrorism, weather forecasting, traffic awareness, etc. These are only a few features that make virtual reality so useful in teaching [14–16].

12.2.1 BETTER PLACE SENSE

Students typically desire to experience something when they read about it. With VR, you don't just talk or illustrate books. You can examine the topic and look at how things are put together.

The sensation that VR gives pupils presence allows them to study a topic by living it. It's easy to forget that VR experiences are not true – a body believes that it is in a different location. The mind is engaged in this emotion in an outstanding way.

Research on the educational utilisation of visually rich 3D virtual environments offers convincing proof of its active and immersive learning capacity [17]. A number of educational initiatives have been implemented, both as an additive to regular school and as a principal medium for remote learning [17–19]. Since 3D virtual environments are relatively new, it is important to know how they may best promote learning. There is a shortage of educational research on 3D virtual environments and their impact on learning results. In recent years, researchers have advanced in virtual worlds, but much more has to be done [20,21]. Research related to the design of 3D environments has found various problems that could have an influence on learning. Some of the challenges with 3D virtual learning environments (VLEs) are isolation, disorientation, navigational difficulty and how to get there [22]. It is difficult for people to recognise and access a specific area, to locate previously visited locations and to build a cognitive environmental map. This is worsened by the virtual environment that is employed as a medium for e-learning and even more so [17]. Another important topic is learning space models in virtual worlds ranging from reproductions of real-life buildings and spaces with the appearance and feel of a true campus to extremities of imagination and fantasy [23–25]. [4] rightly conclude that study into what forms of virtual contexts may be supported by learning and socialising with a broad variety of options is necessary to examine. However, little research has been published on 3D learning environments' design and evaluation. There are therefore few research or recommendations to educate them if institutions are looking to establish learning spaces in 3D virtual worlds. Researchers have in the past offered strategies and models for the development of virtual worlds to help detect and reduce disorientation. This includes using environmental indicators by overlaying semi-transparent virtual maps and grids across the landscape [26], by setting environmental architecture elements such as landmarks and signage [17,27,22]. Some research has been conducted on the influence of location on student learning in 3D virtual environments. The virtual design studios enabled the students to exhibit, discuss and create designs for our websites. The children were pleased in their response to learning. Same researchers later established a location in the classroom and utilised it to teach a web design course surrounded by student galleries. By analysing the talks held in the virtual field, the students were given an approach and a context to the topic through the virtual space.

12.2.2 Scaling the Experience of Learning

Virtual reality gives schools the capacity to extend their curricula to make learning more dynamic and attractive. Students may access a full science laboratory, a historical event or geographies from around the world with one device. Virtual worlds assist in alleviating distractions and promote numerous forms of learning due to their immersive nature. In addition to reading about it, students can learn through seeing or doing things. It is understandably far more interesting to give kids the opportunity to sit on the edge of a volcano, stroll through a group of dinosaurs and to build an aeroplane engine to learn the topic than standard teaching approaches. Technologies like scientific laboratories are incredible – they allow students to grasp how things function from practise.

However, these technologies are costly and very hard to measure. The amount of things they can perform is similarly restricted.

The smooth classroom paradigm and hybrid teaching programmes give us no physical location to play in education in virtual reality. VR enables us to bridge the gap globally between students and teachers. Distance learning solutions, such as the Provided Learning Feature on the Heizenrader XR Education Platform (FDLP), allow students and teachers to occupy a shared virtual space in multiple places (different schools and countries) to discuss courses and topics. VR also allows students to converse, using avatars and mapping facial expressions, to build social learning experiences.

12.2.3 LEARN BY DOING

It is widely accepted that people learn best by doing things yet, when you monitor present learning, you may nevertheless notice how little learning happens. Students focus on reading instructions instead of practising them.

In addition to enabling educators to generate material in a classroom that is consistent with the class curriculum, the VR educational content also allows them to study deeper and monitor and analyse the development of students at an individual level. VR enables professors to immediately recognise and aid pupils with the topic. VR enables professors to immediately recognise and aid pupils with the topic. Teachers can choose to have their own VR material created by students, which encourages pupils to take on a higher degree of responsibility and therefore increase trust.

VR supplies the teaching with an experience anchor. Students are eager to learn VR training for themselves. Instead of passive reading, students have the opportunity to learn.

12.2.4 EMOTIONAL REACTION

In order to develop memories, visceral reactions to what we experience are crucial. VR makes it possible for students to become involved all the time, making their experiences unforgettable.

Classical emotional evoking approaches for examining their impact on in-laboratory cognitive processing, for instance in presenting emotional photos or videos, may not be particularly successful, as triggering targeting emotions is unreliable. The introduction of more immersive, complex multi-sensory display technologies such as VR headsets can overcome this obstacle to produce an emotional stimulation. In the study, scientists used a VR gadget with a load of emotions to stimulate participants' anger and anxiety to examine their results in decision making. The results of this study are in accordance with our assumptions that the influence of target emotions on decision making was greater in connection to VR conditions than in the desktop situation. In comparison with anger conditions, under addition, individuals took longer time per pump (on the BART), demonstrating that in the fear condition the participant is risk-averse. Although the results are supported by other research on the effects of fear and anger on decision making, when the VR system was employed as the medium of elicitation, these reactions are exacerbated. This shows that usage of VR as a medium of emotional excitement can be more successful than traditional approaches for studies on emotional impact on decision making. In addition, preliminary results from this study are based on the PANAS-X answers and imply that VR may be utilised to cause anxiety and panic more than traditional approaches.

12.2.5 DEVELOP CREATIVITY

It is vital to have virtual reality in education, not just for content consumption, but also for content production. By giving wonderful tools, such as Tilt Brush, you help children improve their creativity.

As engineers enhance the software and technology behind the virtual reality, we will continue to extend the ways in which we use this flourishing technology. And although many of the possibilities of VR's gaming, education and entertainment applications have been published, the idea of being utilised to stimulate creativity is rather disregarded by scholars.

Numerous scientists and developers are thus researching methods in which VR uses its capability to liberate its users' creative potential. A recent study in China has showed, for instance, that VR users who had to break down virtual walls had higher creative performance in comparison to those who did not experience any difficulties in the simulation. These users were also found to be more persistent and flexible and suggested that virtual experience in overcoming hurdles boosted conceptual processing.

While additional research is needed to corroborate the conclusions of the study, numerous developers have created sophisticated tools to maximise the creativity of their customers. Here are some of the most popular systems accessible for the development of creativity.

12.2.6 VISUAL LEARNING

In 2016, Samsung's student and instructor survey showed that few have truly harnessed the technology when utilising VR as a training tool:

1. While 85% of instructors believed that VR in the classroom was valuable, currently just 2% used it in any way.
2. The idea of wearing a VR headgear was exciting for 93% of both students and teachers.

However, with progress made in VR hardware and software constantly reducing the cost of setup and deployment, VR for teaching and training is becoming more common on a daily basis. And why it's so exciting is not surprising. There is something in technology for anyone with its multi-layered approach to learning.

12.2.6.1 VR for Visual Learners

VR offers distinct benefits for visual apprentices who may see a process or method in numerous ways, freely study a virtual object's behaviours and subtleties and take up visual material with an immersive 3D experience.

For visual learners in a number of professions, some natural applications may include:

- Health care: complication of surgery or similarly vital but less difficult medical work, such as day-to-day care.
- Public safety: examination of containers, labels or ships to detect the type of dangerous matter that might occur as a car accident is reported in the first responders.

12.2.6.2 VR for Auditory Learners

VR can also assist trainees in industry who need to make use of aural communication both as part of everyday work and as an ideal manner of learning including:

- Energy and public services: gas trainees can communicate about an alleged natural gas leak in MUTIL VRTroaching via VoIP (Voiceover Internet Protocol) and also transmit and receive spoken information from/to (actual or simulated) supervisors.
- General industry: trainees can hear ambient noise, special warnings or noises from equipment that better imitate fast-paced operations and keep them involved in their activities as they execute a normal maintenance method.

Many people are visual learners – for this group of students, VR actually is useful. In fact, pupils see what they are learning rather than reading about topics. It's easy to understand that they are able to envision complex operations or procedures.

Google Expedition: A virtual reality education platform that allows you to conduct or join virtual adventures across the world. Google Expedition lets educators on field excursions access virtual realities in their classrooms. The art gallery or the museum using mobile devices and VR viewers allow a virtual tour for teachers and pupils. Here you may install the Google Expedition app from the play shop.

FotonVR: It is a tool for improving the world's comprehension and opening the doors to the current schooling system. The VR content collection provides future scientists with a realistic learning experience for education, academics, universities, higher education institutions, studies of talents, etc. Using VR technology, you may do field trips with various instruments. We may encounter a fresh and excellent learning technique here rather than regular schools. Start establishing a scientific career for your child or student or give it the ability to grasp any scientific idea.

12.3 NEW TECHNOLOGY IS AVAILABLE TO USERS

When you're thinking about VR technology, the first concept in anyone's head is an entertainment experience. VR is seen by many designers as a game industry expansion. It is true that VR has been devoted to gaming in history, but things change.

The demand for education is more than what gaming content is – 63.9% vs 61% – according to a recent poll by Greenlight VR.

12.3.1 EDUCATIONAL CATEGORIES OF VR

What can we do in education about virtual reality? The reply is nearly everywhere. VR presents an endless array of possibilities for humans. Here are some experiences with VR that you can design.

12.3.1.1 Virtual Fields Trips

VR technology may be utilised to involve pupils in geography, history or literature by delivering a profound feeling of time and location. Just imagine geographical classes where you may visit any area in the world – it's a far more educational experience than just reading about it.

Google Expeditions is a fantastic example of an app that offers such an experience. Expedition is a field trip library offered to smartphone users on a regular basis. Each journey includes vistas of VR, including the journeys from China's Great Wall to Mars. People throughout the world can visit sites that cannot be visited personally.

In hundreds of schools across the world, Google has led this app. Google took over 1 million students on expeditions in 11 countries. The experiment was tremendously successful.

Of course, virtual reality never substitutes for true travels and excursions or should replace them. However, VR permits experiences that would otherwise not be feasible.

12.3.2 HIGH-TECH TRAINING

VR is an ideal option when it comes to high-tech training, such as the military and medical industries. The main hurdle for medical students to study anatomy, for example, is to understand the body in three dimensions and how the different systems work together. VR training can help overcome this problem.

One excellent example is the system for VR in the Opava City, Czech Republic utilised by the Mendel Grammar School to help pupils study the anatomy of the eye during biology courses. A Leap Motion controller and Oculus Rift headsets were specifically designed for unique learning methods by the project team.

12.3.3 INTERNSHIPS

An important aspect of the learning process is exposure to other vocations. We have dreamed about what we want to be from our early youth, and the professional in our life typically encourage our goals. Often via internships we obtain this insight.

The potential to widen pupils' exposure to occupations is a further benefit from virtual reality in education. It strengthens the ability of people to think in the shoes of others. Profession explorations teach how to work in an area – students may explore a day in the career of anyone, meet the person who studies and know what they love – or do not like – about their jobs. As a consequence, kids get to know the experience.

12.3.4 GROUP LEARNING

Some of the most important knowledge we acquire is not the ideas of the professors, but the university and debate. VR training offers pupils the opportunity to socialise learning sessions. People may speak, synthesise and learn from one other via avatars and monitored face expressions.

12.3.4.1 Distance Learning

VR helps us to bridge the gap between students and teachers. VR teachers and students may telephone the VR environment with their own digital representations and guide students through activities in a room.

12.3.4.2 Challenges in Design for VR Training Experiences

The future is clearly very interesting and potentially full with virtual reality in education. This strong technology is just beginning and VR design is full of obstacles we should be prepared to overcome.

12.4 FIVE MAIN CHARACTERISTICS OF A VR EXPERIENCE

The following qualities are expected for VR education apps:

a. Immersive. Designers should try to make people believe what they are experiencing. If you are developing a historical app, for example, make history come alive for kids.
b. User-friendly. Remove the requirement to engage with a VR 3D design programme with unique abilities.
c. Meaningful. This is extremely important for pupils. You can't make a great VR study experience without a good narrative. That's why it is so crucial to promote the art of storytelling. Stories simply provide the best venue for communicating messages that are not only heard and understood but also motivate action.
d. Adaptable. "I never educate my schoolchildren, I just try to provide them the conditions in which they can learn," Albert Einstein stated once. Students should be able to explore VR activities at their own speed. The app should ensure full control of difficulty levels. The creation and utilisation of this knowledge to create VR goods to make effective learning possible for the students should be established.
e. Measurable. Measured impact should be achieved with each teaching instrument. Teachers should be able to monitor the learning measurements in order to measure a subject's results. It is important that the right metrics be chosen and that the criteria to assess success and failure are clearly established when building VR training experiences.

12.5 DEVELOPMENT OF A NEW ROLE FOR VR EDUCATORS

The move from analogue to digital education will affect the way education looks. From material delivery to content facilitation, the job of a teacher will be altered.

Teachers are focused instead on imparting ready knowledge on building the necessary circumstances for exploration.

12.5.1 KNOWLEDGE OF IMMERSIVE EXPERIENCES

It is obvious that we are at the beginning of VR – and it will continue to change. The technology will nevertheless continue to push the limits on how immersive VR can become. In the future years, we will even witness progress in eye tracking. What we consider to be an immersive experience today may not be in the not-too-distant future.

12.5.2 MAKING VR ACCESSIBLE

VR has a high price point for many users, which is a big hurdle. It's clear that Google could reach many of these children, because the hardware was accessible when we analyse the success of Google. In order to provide VR training to a majority of users and convert it into outstanding teaching tools, the production of VR experience is important for device buyers. It is important. A phone that we carry in our pockets, such as Google Cardboard or Samsung Gear VR, combined with Rs. 1000–1500 in headgear should provide our children with an excellent VR experience.

12.6 OPEN NEW HORIZONS WITH VR EDUCATION – FUTURE

Virtual reality is on the horizon in education and will surely transform the world as we know it. The classrooms in the 21st century will be high-tech learning environments, with VR technology significantly improving student engagement and learning. VR experiences will inspire a whole new generation of young and brilliant students who are ready to innovate and change the world.

At the same time, a teacher's choice to adopt such technology in classrooms is no longer the technology, but the next key aspect of education. The international goal should be to offer accessible and cheap knowledge for everybody on earth. Further information about the design of different, new technologies may be found here.

REFERENCES

[1] F. P. Brooks, "What's real about virtual reality?," *IEEE Comput. Graph. Appl.*, vol. 19, no. 6, pp. 16–27, 1999, https://doi.org/10.1109/38.799723.
[2] J. Psotka, "Immersive training systems: Virtual reality and education and training," *Instr. Sci.*, vol. 23, no. 5, pp. 405–431, 1995, https://doi.org/10.1007/BF00896880.
[3] D. Allison and L. F. Hodges, "Virtual reality for education?," in *Proceedings of the ACM Symposium on Virtual Reality Software and Technology*, pp. 160–165, 2000, https://doi.org/10.1145/502390.502420.
[4] S. Kavanagh, A. Luxton-Reilly, B. Wuensche, and B. Plimmer, "A systematic review of virtual reality in education," *Themes Sci. Technol. Educ.*, vol. 10, no. 2, pp. 85–119, Dec. 2017, [Online]. Available: https://www.learntechlib.org/p/182115.

[5] R. B. Loftin, M. Engleberg, and R. Benedetti, "Applying virtual reality in education: A prototypical virtual physics laboratory," in *Proceedings of 1993 IEEE Research Properties in Virtual Reality Symposium*, pp. 67–74, 1993, https://doi.org/10.1109/VRAIS.1993.378261.

[6] Y. Lau, K. T. Lau, and Y.-T. Chow, "Outcome-based education in Hong Kong sub-degree institutions," *Int. J. Innov. Educ.*, vol. 4, no. 4, pp. 280–296, 2017, https://doi.org/10.1504/IJIIE.2017.091504.

[7] E. Çetin and S. Özdemir, "Children's problem-solving with programming activities: A case study with small basic," *Int. J. Innov. Educ.*, vol. 4, no. 4, pp. 264–279, 2017, https://doi.org/10.1504/IJIIE.2017.091503.

[8] H. Alharbi and K. Sandhu, "Managerial staff perceptions on the e-learning re-commender system: A case of Saudi Arabia," *Int. J. Innov. Educ.*, vol. 4, no. 4, pp. 249–263, 2017, https://doi.org/10.1504/IJIIE.2017.091486.

[9] R. Klavir and J. Goldenberg, "Policy-shaping graduate follow-up studies: The case of a program for excellent students in colleges of education," *Int. J. Innov. Educ.*, vol. 4, no. 4, pp. 227–248, 2017, https://doi.org/10.1504/IJIIE.2017.091484.

[10] E. Hu-Au and J. J. Lee, "Virtual reality in education: A tool for learning in the experience age," *Int. J. Innov. Educ.*, vol. 4, no. 4, pp. 215–226, 2017, https://doi.org/10.1504/IJIIE.2017.091481.

[11] D. L. Andolsek, "Virtual reality in education and training," *Int. J. Instr. Media*, vol. 22, no. 2, p. 145, 1995, [Online]. Available: https://www.learntechlib.org/p/85408.

[12] A. Tal and B. Wansink, "Turning Virtual reality into reality: A checklist to ensure virtual reality studies of eating behavior and physical activity parallel the real world," *J. Diabetes Sci. Technol.*, vol. 5, no. 2, pp. 239–244, 2011, https://doi.org/10.1177/193229681100500206.

[13] P. S. Bordnick, B. L. Carter, and A. C. Traylor, "What virtual reality research in addictions can tell us about the future of obesity assessment and treatment," *J. Diabetes Sci. Technol.*, vol. 5, no. 2, pp. 265–271, 2011, https://doi.org/10.1177/193229681100500210.

[14] G. Makransky and L. Lilleholt, "A structural equation modeling investigation of the emotional value of immersive virtual reality in education," *Educ. Technol. Res. Dev.*, vol. 66, no. 5, pp. 1141–1164, 2018, https://doi.org/10.1007/s11423-018-9581-2.

[15] M. Němec, R. Fasuga, J. Trubač, and J. Kratochvíl, "Using virtual reality in education," in *2017 15th International Conference on Emerging eLearning Technologies and Applications (ICETA)*, 2017, pp. 1–6, https://doi.org/10.1109/ICETA.2017.8102514.

[16] G. Schröder, M. Thiele, and W. Lehner, "Setting goals and choosing metrics for recommender system evaluations," *CEUR Workshop Proceedings*, vol. 811. pp. 78–85, 2011.

[17] M. D. Dickey, "Teaching in 3D: Pedagogical affordances and constraints of 3D virtual worlds for synchronous distance learning," *Distance Educ.*, vol. 24, no. 1, pp. 105–121, 2003, https://doi.org/10.1080/01587910303047.

[18] S. A. Barab, K. E. Hay, M. Barnett, and K. Squire, "Constructing virtual worlds: Tracing the historical development of learner practices," *Cogn. Instr.*, vol. 19, no. 1, pp. 47–94, 2001, https://doi.org/10.1207/S1532690XCI1901_2.

[19] R. Kumar and A. Anand, "Internet banking system & security analysis," *Int. J. Eng. Comput. Sci.*, vol. 6, no. 6, pp. 2319–7242, 2017, https://doi.org/10.18535/ijecs/v6i4.43.

[20] M. H. Davis and M. G. Gaskell, "A complementary systems account of word learning: neural and behavioural evidence," *Philos. Trans. R. Soc. B Biol. Sci.*, vol. 364, no. 1536, pp. 3773–3800, 2009, https://doi.org/10.1098/rstb.2009.0111.

[21] W. S. Bainbridge, "The scientific research potential of virtual worlds," *Science*, vol. 317, no. 5837, pp. 472–476, 2007, https://doi.org/10.1126/science.1146930.

[22] V. H. Wright, G. E. Marsh, and M. T. Miller, "A critical comparison of graduate student satisfaction in asynchronous and synchronous course instruction," *Plan. Chang.*, vol. 31, no. 1, p. 107, 2000, [Online]. Available: https://www.learntechlib.org/p/96312.

[23] T. Hew-Butler *et al.*, "Maintenance of plasma volume and serum sodium concentration despite body weight loss in ironman triathletes," *Clin. J. Sport Med.*, vol. 17, no. 2, 2007, [Online]. Available: https://journals.lww.com/cjsportsmed/Fulltext/2007/03000/Maintenance_of_Plasma_Volume_and_Serum_Sodium.5.aspx.

[24] M. Chakraborty and K. M. A. Khan, "A case study on the importance of human resource information system in the healthcare sector of a corporate hospital in India," in *2019 International Conference on Digitization (ICD)*, pp. 115–117, 2019, https://doi.org/10.1109/ICD47981.2019.9105864.

[25] P. Pandey, S. Mishra, P. Rai, and A. Anand, "Social engineering and exploit development," 2019, [Online]. Available: www.ijsrcsams.com.

[26] R. P. Darken and J. L. Sibert, "Navigating large virtual spaces," *Int. J. Human–Computer Interact.*, vol. 8, no. 1, pp. 49–71, 1996, https://doi.org/10.1080/10447319609526140.

[27] L. E. Thomas, L. A. Chariot, and J. T. Stanley, "Computer-aided analysis of oriented crystallites by diffraction pattern simulation and tilt-stage control in a TEM," *Microsc. Microanal.*, vol. 3, no. S2, pp. 1015–1016, 1997, https://doi.org/10.1017/S143192760001196X.

13 Concentrated Gaze Base Interaction for Decision Making Using Human-Machine Interface

B.G.D.A. Madhusanka[1], Sureswaran Ramadass[1], Premkumar Rajagopal[1] and H.M.K.K.M.B. Herath[2]

[1]School of Science and Engineering, Malaysia University of Science and Technology (MUST), Petaling Jaya, Malaysia

[2]Department of Mechanical Engineering, Faculty of Engineering Technology, The Open University of Sri Lanka, Nugegoda, Sri Lanka

13.1 INTRODUCTION

In human-computer interaction (HCI), an increasing number of academics have noted that computer science and technology continues to progress while HCI is evolving. Human-computer interaction technology [1–2] has evolved from a one-stop all-keyboard to modern multimedia connections. Thus, HCI is becoming prevalent via many media, including look tracking and speech recognition, in conjunction with human input. Moreover, 80–90% of external information is received utilizing the eyes. Thus, by monitoring the eye gaze, people may acquire visual comprehension of information [3–4]. The usage of computer vision technologies has grown more and more in fields including medical, gaze tracking [5], production testing [6], human-machine interactions [7–8] and military aviation [9].

The visual system may provide more than 80% of the information. An eye is crucial to convey the emotional condition, needs, cognitive processes and other aspects [10]. Furthermore, what a person is searching for may define the behaviours or intentions. Gaze tracking is thus a component of the HCI discipline. The visual line of sight of a user, their mental intents and their conduct may be understood as a gaze tracking device for collecting changes [11]. This data may reflect the link between the information of the user's eye movements and the choice of thought and knowledge. They provide a theoretical and practical basis for psychological and ophthalmological studies [12]. Companies and academic organizations have developed high-precision gaze-tracking systems based on professional equipment [13]. These technologies are

DOI: 10.1201/9781003196686-13

used, for example, in medical environments, assisted driving systems and schools. However, commercial gaze tracking devices are costly, limiting their use [14].

While most gaze tracking systems on the market are based on active infrared or stereo camera sources that need specialized hardware support and intricate hardware parameters and position calibration [15], it is thus not suitable in the educational and advertising sectors. Allow gaze tracking to be utilized more broadly. If you use one monocular camera [16], a system may be developed for gaze tracking without specialized or intrusive equipment.

Gaze tracking is a process through which the gaze point in space or visual axis is determined. Visual scanning pattern analyzes and gaze tracking systems are utilized in the HCI. The eye view may be used in HCI to replace conventional input devices like a mouse pointer [17] as advanced input computer input [18]. The display may also be controlled interactively by the eye gaze on the screen [19]. Given that visual scanning patterns are strongly associated with the centre of attention of the individual, cognitive scientists investigate human cognitive processes by using the gaze-tracking system [20]. Overall, the video-based estimated methodology for the eye gaze may be divided into two groups: 2D gaze-based mapping [21] and 3D gaze-estimated methods [22], which estimate the participants' 3D visual axis. An eye-tracking survey is available in [23]. The 3D techniques have recently become more popular due to their excellent precision under free head movement. However, more sophisticated 3D gaze estimates need a calibration process to estimate their eye characteristics for everyone.

The 2D mapping technique, using conventional gaze estimation, learns from the 2D functions a polynomial mapping, e.g., the 2D pupil glint vector [24] and 2D eye [25] to the gaze on the screen. There are two typical disadvantages of the 2D mapping method. The user first has to carry out a complicated experiment to calibrate the mapping parameters to understand the mapping function. The patient must gaze at nine uniformly dispersed spots on the screen or gaze at 12 points to improve their accuracy in the calibration process [26]. Secondly, since the derived 2D image characteristics substantially vary with head movement, the view mapping function is sensitive to head motion. Zou et al. [27] revealed how gaze tracking systems collapse when the head moves away from its initial calibration location. The user must thus hold their head abnormally motionless to perform well. Methods for managing head position changes with the neural network [28] or SVM [29] have also been suggested. However, these approaches only consider the translation of the head in the plane [30], or stereo cameras are needed for the 3D eye location [31].

A 2D appearance-based estimation of a gaze without calibration has proposed by Majaranta et al. [32]. You experiment while you view a picture or movie on the screen. Given the picture's saliency map, look points are collected and utilized as training data to train a mapping function between the image and the viewing point (Gaussian Process Regresser). However, the system's accuracy is minimal compared to state-of-the-art techniques [33] because of the significant incertitude of the saliency plane. Because the 2D mapping procedure does not consider the head position, the chin station must correct the head. Intrusive gaze tracking technology [34] or non-invasive (head-free) [35] may be categorized eye-tracking technology may be categorized as 2D gaze tracking [36,37] and 3D gaze tracking [34], depending on the difference in gaze direction estimations in dimension. The techniques of tracking the gaze include Limbus Tracking [38], Pupil Tracking [39], Pupil-glint Vector [40], Purkinje Image [41], etc., for various gaze tracking systems.

The map function between view locations and the target plane or areas of interest has been determined for 2D gaze estimation techniques first [42]. The resolved mapping function is then used to compute the point of view on targets or areas. A human eyeball model is used to estimate the absolute location of eyeballs in the testing area for the 3D gaze estimation techniques. The 3D gaze is computed accordingly to get the precise gazing position or targets of fixation of human eyes in space [43].

The most often-used calculation techniques may be classified into two categories in traditional two-dimensional estimation approaches: linear (direct least square) [44] and non-linear (generic artificial neural network) [45–46] To compute the mapping function between the calibration markers and the associated pupil-glint vectors, Hassoumi et al. [47] use the smallest squares. A sequence of polynomials comprises overdetermined linear equations in the second order for mapping function. The number of polynomials varies according to the number of markers [48]. A calibration procedure is used to identify the respective calibration coordinates, pupils and glint centres. The pupil-glint vector is determined by the removal of the coordinates of the pupil and glint centre. Wöhle et al. [49] offer a technique of adaptive calibration. For error correction, a second-time calibration is used. The modeling function is based on a higher-order polynomial transformation utilizing medium square error criterion. The results of the single calibration indicate that the precise gaze estimations increase in an enhanced order of polynomials. However, the accuracy of the gaze monitoring scheme does not improve with the more polynomial order due to head motion components, a large number of calibration markers and pupil-glint vector calculations via experimental analyses by Jha et al. [50]. The second linear polynomial is the most popular technique of linear regression, with fewer calibration markers and better mapping resolution approximation [51].

The main aim of this chapter is to estimate the gaze on a monitor attached to the head of the human eye. Mapping function from plane to plane between gaze locations and fastening objectives. Therefore, a new technique for calculating the direction of the gaze-based on a pupil-vector [52–53] is suggested. The combination of a 2D eye model is used for this procedure. First, the suggested approach assesses the likelihood of the eye parameter and the eye look, unlike conventional 2D methods, which estimate the eye and eye parameter deterministically. Thus, ambiguity in the system may be managed better. Secondly, we developed an incremental learning technique to progressively increase the estimated estimation if the subject uses the system organically. Explicit calibration processes or goals for calibration are utilized in both instances. The experimental findings indicate that the average accuracy of our systems for various individuals is above 86%.

13.2 PROPOSED METHOD FOR GAZE ESTIMATION

In this chapter, a new technique is provided for estimating the pupil's vector according to the particular features of linear and non-linear regression. To solve the map function between the pupil vector and the point of gaze and determine the direction of gaze, a robust and efficient artificial neural network (ANN) was established. The flow method for the gaze estimate is illustrated in Figure 13.1. First, a camera connected to the laptop screen collects the eye images of participants while gazing at the calibration marker on the screen.

After calibrating parameters, we used bilateral filters to smooth the image and preserve the edges of the image content, and it was suitable for retaining the pupil's features. Then, equalize the histogram to enhance the contrast. Then, approximate the average intensity within the pupil by the surrounding regions by filling them with the average intensity. Then, apply global thresholding by considering the average intensity within the pupil and invert the image to highlight the pupil blob. Nonetheless, a few other blobs, which are as dark as the pupil region, such as eyelashes, eyelids and shades, remain in the binary image. To distinguish the actual pupil region from the noisy blobs, morphological operations for the noise removal are performed. Among the remaining candidate blobs are determined the final pupil by considering the blobs' shape, size and location. Its centre of gravity is then used as the pupil centre feature.

A Canny edge detector and a Hough circular transformation are used to identify the pupillary iris border. The fusion of the profile with the mask methods is utilized to identify the borders of the iris for pupil detection and the circular Hough transformation. For the outer iris border, the gradients are inclined vertically. Based on threshold levels, the disparity between the pupil and iris centres and the radii is obedient. Then the accurate detection is evaluated. Second, the pupil vector is computed as the input of the proposed ANN by removing the iris and the pupil centre coordinates. Third, the mapping function between pupil vectors and associated gaze locations is calculated using a three-layer ANN (input layer, hidden layer and output layer). The gaze spots on the screen may finally be calculated based on the mapping function.

The ANN is used to build the mapping relations (x, y) representing the pupil estimation coordinates of the left and right eyes to estimate the left and right visual axes and gaze point coordinates. The NN with one hidden layer was chosen because of its advanced capability of modelling highly non-linear relationships. In this study, pupil position (x *and* y component) of both eyes are selected as the input features for estimating each eye's gaze-point coordinates.

Before the NNs can perform the estimation, they need to be trained with the data collected in a calibration process. The hidden layer in the NN model is an essential component that can significantly affect the estimation of the gaze vector. Investigate its effect and find a better NN configuration; hidden layers with different hidden units were tested. The changes of the hidden units first affect the estimation of the gaze vector, which eventually affects the estimation accuracy of the 2D gaze. Thus, for the optimal estimation, the number of hidden units can be chosen to be five. Though the error of the gaze vector increases, it does not affect the final 2D gaze estimation because of the gaze vector method characteristic that, when the gaze vector has a small error. Figure 13.2 illustrates an enhanced NN scheme consisting of three layers, including input, hidden layer and a layer of output. The pupil vectors of both eyes enter elements as the neural network inputs. The gaze point coordinates

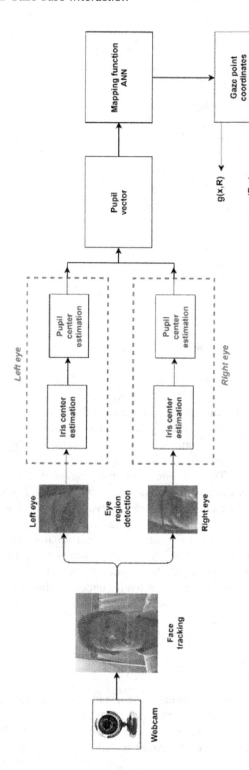

FIGURE 13.1 Flow-Process of Gaze Direction Estimation.

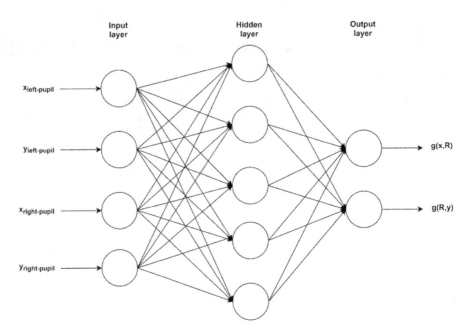

FIGURE 13.2   Scheme Framework of Three-layer NN.

output elements will be calculated as the NN output. There are four, two and five nodes correspondingly to input, output and buried levels.

     Video-based vision tracking systems are included in a camera with a user interface for user eye view tracks. Figure 13.3 shows the typical setup for the eye-gaze monitoring. User calibration, visual and eye videos are typical phases of passive video-based eye-striking, visual recognition and mapping with the gaze coordinates on the screen. The direction of the eye is determined based on both the pupil and the iris centre. Gaze estimation is the technique through which a person's 2D view line may be estimated or tracked mostly while seeing. It has computed the relative gaze motion between the pupil centre and the iris position. The gaze tracking user interface may be active or passive, individual or multi-modal [54]. The gaze of the user may be detected for an active user interface to be activated. Gaze data may be used as some input. An interface without a command is a passive interface in which eye-gaze data are collected for understanding users' interests and attention. The gaze is the single variable for entries for single mode eye-tracking interfaces. In contrast, a multi-modal input combines mouse, keyboard, touch or blink input with gaze input.

13.2.1   Calibration

Figure 13.4 depicts anatomy and model generalized of the human eye. The criteria for the eye are typically the centre of the pupil, the centre of the cornea, visually and optically [55,56]. The back of the eyeball is termed the *retina*, while in the center of the retina the fovea has the greatest visual sensitivity. The visual axis is the line

FIGURE 13.3 Schematic Diagram of a Typical Gaze Tracking System.

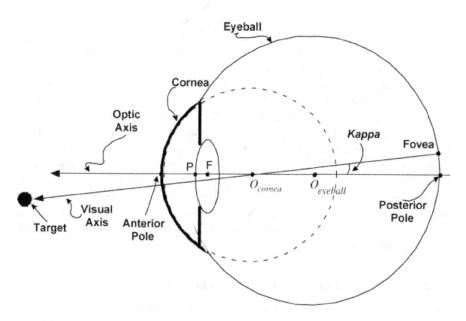

FIGURE 13.4 Model of a Human Eyeball.

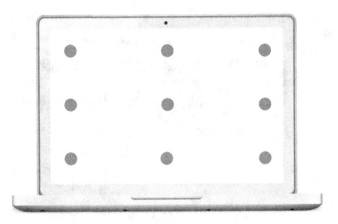

FIGURE 13.5   Calibration Screen with Nine Target Points.

between the center of the fovea. The optical axis is the line between the center and center of the pupil. The direction of the gaze varies from the optic axis with the axis of visual impression. This variance is called a kappa angle and is about 5 degrees. Kappa angle panel and tilt are calculated on a gaze estimate for each user. Thus, everyone needs a calibration process, which must be carried out at the beginning of eye monitoring to acquire a visual axis and kappa angle.

The user is given a series of targets dispersed over the front calibration, as illustrated in Figure 13.5, and the user is requested to gaze at them for some time [57]. The webcam collects the different eye locations of each destination point, and then the tracker learns this mapping function by mapping it to the relevant eye coordinates. The number of target points, the user time for each dot and the kind of mapping method employed thus vary in the calibration procedures.

In general, the calculation of at least 3, 6 and 10 polynomial coefficients are necessary if a first, second and third linear polynomial are used for calibration, meaning that a minimum of 3, 6 and 10 calibration markers are necessary. If too many calibration markers are necessary, unimportant inputs may be deleted according to the primary component analysis to decrease the number of coefficients for polynomial resolution. Generally, four and five calibration target point models are most often used for calculating the first order based on an overall assessment of the real-time quality and precision of the gaze-monitoring system. By contrast, 6 and 9 target point calibration models have been used most often for calculating the second order [58]. This chapter thus examines the mapping function model of nine calibration target locations.

13.2.2   ESTIMATION OF GAZE-TRACKING ACCURACY

Visual stimulation in a set of objectives or scenarios is offered when a user looks at a user interface in a typical visual monitoring operation on a computer screen. The accuracy of the gaze-tracking procedure is evaluated as an average difference between the natural stimulus and the gaze location. Concerning the gaze, precision

tracking is assessed with pixel distance and distance. The computations given below
are these accuracy estimates.

Calibration techniques are designed to aid the systems incorrectly calculating the
Point of Gaze (PoG). Therefore, a decision about the number of calibration points is
critical. An effective calibration algorithm must have as many numbers as possible
of calibration points, making the user familiar with the system. On the other hand, it
should also be simple enough not to cause trouble to the user. It may be helpful to
determine the eye area utilized to scan the laptop screen behind the concept of a
calibration method. In inaccessible mode, the user may remain at a corner point for
an indefinite length of time. This concept enables the system to minimize mistakes
caused by the miscalibration of the gaze. Single computations are given for brevity,
and the same equation applies to the left and right eyes. PoG_{x-left}, PoG_{y-left},
$PoG_{x-right}$, $PoG_{y-right}$ are the measured x, y coordinates of the left and right eye's
PoG. The mean gaze coordinates consider both eyes as PoG_x and PoG_y are the
distance of the eye from the screen in Equations (13.1), (13.2) and (13.3), respectively. The x_{pixel} and y_{pixel} pixels are the pixel shifts in x, and y directions and
offset is the distance between the webcam and lower edge of the display screen in
Equation (13.4) [59].

Gaze-point coordinates:

$$PoG_x = \frac{PoG_{x-left} + PoG_{x-right}}{2} \tag{13.1}$$

$$PoG_y = \frac{PoG_{y-left} + PoG_{y-right}}{2} \tag{13.2}$$

Pixel accuracy:

$$Pixel\ accuracy = \sqrt{(Actual_x - PoG_x)^2 + (Actual_y - PoG_y)^2} \tag{13.3}$$

On-screen distance:

$$On\ screen\ distance = pixel\ size$$

$$\times \sqrt{\left(PoG_x - \frac{x_{pixel}}{2}\right)^2 + \left(y_{pixel} - PoG_y + \frac{offset}{pixel\ size}\right)^2} \tag{13.4}$$

13.3 EYE-GAZE ESTIMATION ALGORITHM

The vector between the iris centre and the pupil centre is mapped using a polynomial transformation function or a geometric eye model utilizing a regression-
based technique to match the gaze locations on the frontal screen. The 2D regression is utilized to evaluate the direction of the gaze, which is given afterwards.

The pupil vector is first computed utilizing the pupil and the iris centre. The second linear gaze-mapping function is given in Equations (13.5) and pupil vector (13.6).

$$x_c = a_0 + \sum_{p=1}^{N} \times \sum_{i=0}^{p} a_{(i,p)} X_e^{p-1} Y_e^i \tag{13.5}$$

$$y_c = b_0 + \sum_{p=1}^{N} \times \sum_{i=0}^{p} b_{(i,p)} X_e^{p-1} Y_e^i \tag{13.6}$$

where $i = 1, 2, 3, ..., N$. N is the number of calibration points, (x_c, y_c) is the co-ordinate of gaze calibration markers on-screen coordinate system and (X_e, Y_e) is the coordinate of pupil vector on the image coordinate system. As conventional linear methods, least squares is utilized to solve the gaze-mapping function shown in Equations (13.5) and (13.6). In addition, the polynomial is optimized through calibration in which a user is asked to gaze at specific fixed points on the frontal screen.

The order and coefficients are then chosen to minimize the mean squared difference (ε) between the estimated and actual screen coordinates (13.7):

$$\varepsilon = (x_c - Ma)^T (x_c - Ma) + (y_c - Mb)^T (y_c - Mb) \tag{13.7}$$

where a and b are the coefficient vectors and M is the transformation matrix given by:

$$a^T = [a_0 \quad a_1 \quad \cdots \quad a_m] \tag{13.8}$$

$$b^T = [b_0 \quad b_1 \quad \cdots \quad b_m] \tag{13.9}$$

$$M = \begin{bmatrix} 1 & X_{e1} & Y_{e1} & \cdots & X_{e1}^n & \cdots & X_{e1}^{n-i} Y_{e1}^i & \cdots & Y_{e1}^n \\ 1 & X_{e2} & Y_{e2} & \cdots & X_{e2}^n & \cdots & X_{e2}^{n-i} Y_{e2}^i & \cdots & Y_{e2}^n \\ \vdots & \vdots & \vdots & \cdots & \vdots & \cdots & \vdots & \cdots & \vdots \\ 1 & X_{eN} & Y_{eN} & \cdots & X_{eN}^n & \cdots & X_{eN}^{n-i} Y_{eN}^i & \cdots & Y_{eN}^n \end{bmatrix} \tag{13.10}$$

where M is the transformation matrix, m the number of coefficients and N the calibration points. The coefficients can be obtained by inverting the matrix M as:

$$a = M^{-1} x_c, \quad b = M^{-1} y_c \tag{13.11}$$

In this chapter, nine points of calibration ($N = 9$) are used and second-order polynomial transformations. A calibration target and mapping function component arrangement play an essential part in establishing the overall correctness of the eye look. In this instance, the higher-order components in the mapping function are primarily used to correct for errors in the predicted direction of the gaze. The higher the polynomial order, the greater the exactness of the computation. However, it is also possible to increase the number of polynomial coefficients to be resolved.

Furthermore, there were also increased numbers of calibration markers. It not only lengthens calibration time but also increases the user load by the heavy calibration procedure. Users are prone to tiredness, which affects the accuracy of the calibration. In addition, the mapping accuracy is improved, and the direction of their gaze is precisely estimated. The ANN based on the direct minimum square regression will resolve a mapping function between pupil vectors and screen calibration points.

The steepest gradient descent technique [60] is the method of training of NN to solve the mapping function in Equations (13.5) and (13.6). First, the connection between the hidden and output layers should be established. Next, the minimum square regression direct solution constraint specifies NN's error costing and ongoing learning rule. According to gaze estimation, the Euclid standard is chosen for a low-cost function in the same form as an error fixation criterion for the minimum squares regression, as specified in the equation according to gaze estimation (13.7).

13.4 EXPERIMENTAL SYSTEM AND RESULTS

The experiment focused on detecting visual attention and intention, which are essential elements of the entire framework. As illustrated in Figure 13.6, the home care situation was replicated by a feedback scenario presented to the user. A kitchen with visible items was shown in the scene picture. A camera was utilized to track the subjects on the screen.

The participants were sitting at a monitor showing the picture of the kitchen scene during the experiment. Participants are trying to convey their purpose by gazing at such items with their eyes. In addition to the gaze locations, their eye movements were recorded. Support vector machine (SVM) classifying was used to identify their visual attention and recognize the visual objects from the visual data.

FIGURE 13.6 Environment Setup in This Research.

Participants' actual visual attention and intent were reported throughout the experiment and recorded as ground truth to assess the system's effectiveness.

We are compared to the mouse model, one of the most common interaction models and has excellent usability, utilizing natural visual behaviour for care detection. While the mouse mode is not accessible to most handicapped people, it is nevertheless highly efficient for showing the usefulness of the gaze modality because of its ubiquity and excellent useability. The remainder of the mouse configuration was the same as the gaze.

13.4.1 Perception of Human Intentions

The way to deduce the user's intention may be to indirectly convey the intended goal; for example, via non-verbal instructions. However, if implicated user contact is needed, it may result in successful cooperation. People anticipate the intents of others exceptionally well, showing that nonverbal communication may contain inference intentions. The current paper investigates how the desire of the elderly to convey nonverbal signs and indirect indications that the user implicitly gives while carrying out activities for quicker and natural engagement may be used. This research offers instructions to control the eye and analyse the user's intention to deduce everyday tasks. The suggested technique of this study is the SVM classification to inform the identification of human intents. An assessment is intended to examine the intention inferred and generally is performed alone in domestic cases. A questionnaire based on contextual factors is utilized for the method of intention recognition. The caregivers then decide or diagnose based on a method for identification of intentions.

Four deliberate items have been empirically detected in the kitchen scenario. These four items have been deliberately created as "teacups," "water glasses," "juice glasses" and "soup bowls." The simulated kitchen scene included all handling objects. In this investigation, visual attention and purpose, essential components of the national situation, were detected. The house scenario is modelled on the user feedback scene picture, as illustrated in Figure 13.7.

FIGURE 13.7 Artificial Kitchen Image with Objects.

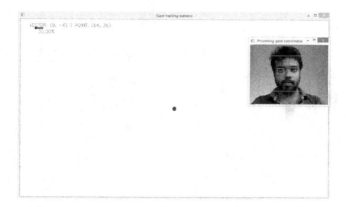

FIGURE 13.8   Calibration Screen.

13.4.2 INTENTION INFERENCE

To evaluate the eye-gaze performance, we recorded a set of videos with 30 participants. For each frame of those videos, the correct gaze direction was considered. Thus, a total of 30 videos were recorded, with different lighting conditions, using glasses or not, and in different positions about a laptop webcam. First, the participants were instructed to calibrate the laptop screen by looking at the nine target calibration points, as shown in Figure 13.8. Then, participants were instructed to start capturing the video by looking at the teacup, glass of water, glass of juice and finally bowl of soup in their eyes, respectively.

The visualization of the intentional gaze and the intentional gaze show different eye-gaze properties. The intentional eye look characteristics are provided to display items in the cooking situation. A person's gaze is more extended during the intentional gaze than the intentional gaze depicted in Figure 13.9 and focuses more on gaze distribution. Consider, in this experiment, that the time to stay is five seconds longer. In the experiment, the SVM classification was then utilized to identify the visual attention of the participants. The user intention is displayed at the top of the screen, where dwelling relatively longer than five seconds. Figure 13.9(a) describes the user intention to teacup, user intention of glass of water is described in Figure 13.9(b). Next, the user intention identifies as glass of juice is described in Figure 13.9(c). Finally, user intention to the bowl of soup is described in Figure 13.9(d), respectively.

The recorded videos are divided by 30 participants with an average processing rate of 30 frames/second regarding their glance at items in the kitchen scenario with around 20 seconds for each participant. The system used the camera under indoor illumination settings throughout this trial. The face of the individual at first was not lit by light from the top or upper corners.

The SVM classifier training gathered 95 sets of positive training data and 70 negative training datasets. These training data sets were utilized for training the classifier, and the total success rate for training was 80.67%. Table 13.1 summarizes more specific training performance.

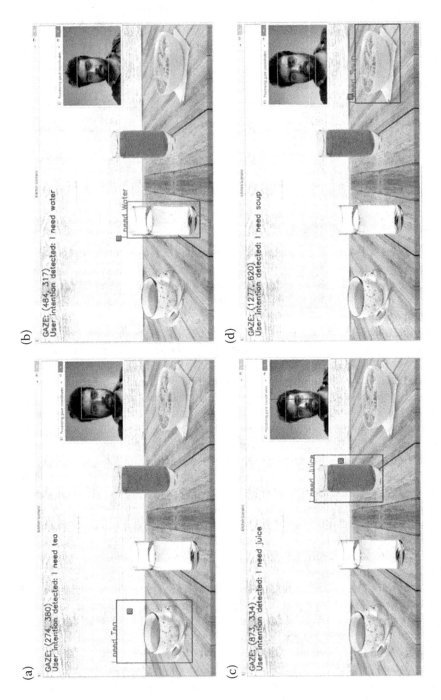

FIGURE 13.9　Intentional Gaze in the Kitchen Environment.

TABLE 13.1

SVM Classifier for Object Intention Detection

Training Data Set	Positive Classification (%)	Negative Classification (%)
Positive training data	77.3	22.7
Negative training data	15.2	84.8

Real attention tends to be less efficient than false detections. The classifier developed. In the experiment, the SVM classifier is then utilized to identify the visual attention of each participant. The SVM classifier identifies user intention in this experiment, using gazing data pre-labelled based on visual attentiveness and user eye gazes closer to the chosen item. For performance assessment, two criteria have been used: accurate rate of detection and fake rate. The positive grading rate for the positive training data set is 77.3%, whereas the negative grading rate is 22.7%. The positive grade rate in the negative training data set is 15.2%, and the negative grade rate is 84.8% correctly identified. In the kitchen scenario, the performance of the SVM attention classifier differed considerably across four distinct subjects. The findings showed the SVM classifier and the chosen eye-gaze characteristics were able to identify users' visual attention during normal visual activities.

13.4.3 VISUAL OBJECT INTENTION

Identify a user intention for a kitchen scenario; four intended items have been chosen and input individually into the SVM model for correlation training. Figure 13.10 shows an example of the accuracy of each kind of purpose-depicted item. Every purpose of the 30 participants in the kitchen scenario is shown in the correlation diagram. While some individuals may have seen various things for intention, most participants viewed familiar objects broadly. Most participants have, for example, adequately chosen the purpose of the water glass.

Object intention to the teacup refers to a linear regression analysis of this data reveals a linear relationship. The coefficient of determination, R^2, is about 0.0814, which means that this equation explains 8.14% of the changeability in object intention. This indicates a negative linear relationship of object intention to the teacup. Then, object intention to the glass of water refers to a linear regression analysis of this data reveals a linear relationship. The coefficient of determination, R^2, is about 0.0948, which means that this equation explains 9.48% of the changeability in object intention. This indicates a positive linear relationship of object intention to the glass of water. Next, object intention to the glass of juice refers to a linear regression analysis of this data reveals a linear relationship. The coefficient of determination, R^2, is about 0.083, which means that this equation explains 8.3% of the changeability in object intention. This indicates a positive linear relationship of object intention to the glass of juice. Finally, object intention to the bowl of soup refers to a linear regression analysis of this data reveals a linear relationship. The coefficient of determination, R^2, is about 0.0354, which means that this equation

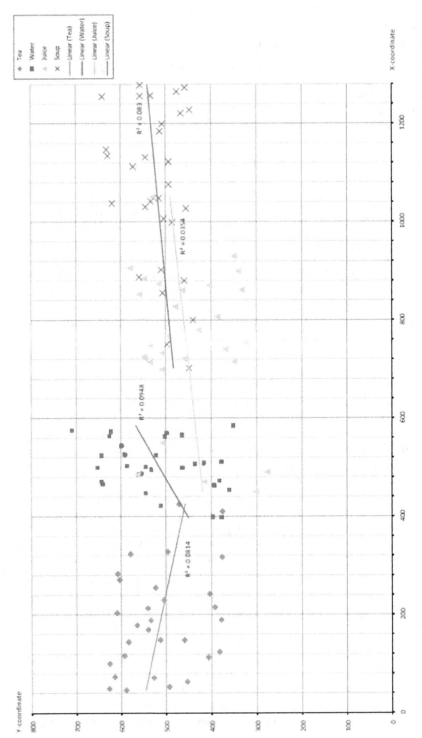

FIGURE 13.10 Correlation Plot between Intention and Objects.

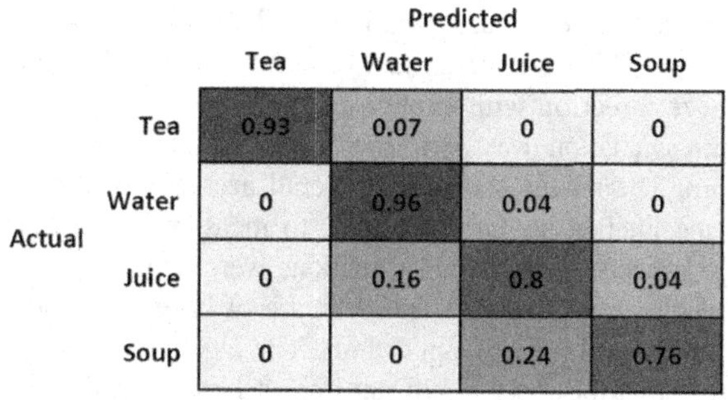

FIGURE 13.11 Confusion Matrix for the Intention to the Objects in the Kitchen Scenario.

TABLE 13.2

Accuracy and Precision of Intention to Each the Object

Gaze Intention	Accuracy (%)	Precision
Teacup	97.5	0.93
Glass of water	93.33	0.97
Glass of juice	89.17	0.80
Bowl of soup	93.33	0.77

explains 3.54% of the changeability in object intention. This indicates a positive linear relationship of object intention to the bowl of soup.

One detailed inference performance of the intention to the objects in the kitchen scenario is illustrated as a confusion matrix shown in Figure 13.11. Again, the horizontal axis is the predicted intention, and the vertical axis is the actual intention.

The precision and accuracy for each type of intention to the objects are summarized in Table 13.2. The precision of the intention to teacup was inferred in 0.93 of the cases. The precision of the intention to glass of water was inferred in 0.97 of the cases. The precision of the intention to glass of juice was inferred in 0.80 of the cases. Finally, the precision of the intention to bowl of soup was inferred in 0.77 of the cases. The overall accuracy rate was 86.7%. Notably, 13.3% of the error occurs for the intention to the object variation in the kitchen scenario. With fewer dominant eye-gaze variations, it is more challenging to characterize the intention from the object aspect. The SVM model has less tolerance for the mistakes of the attention detection classifier.

13.5 CONCLUSION

The estimate of eye gaze is a multidisciplinary area of research and development that has received much interest from academic, industrial and general user groups over recent decades due to the available computer and equipment resources as well as the increasing requirements for HCI technology. This chapter aims to evaluate the human eye's gaze point on display with a fixed head. The mapping function is from plane to plane between gaze locations and fixation goals. In addition, it is suggested that a new 2D gaze technique, based on the pupil vector, computes the motion of the gaze direction with explicit calibration.

The algorithm evaluation was performed using 30 participants through the conventional webcam. The results showed an overall accuracy of 86.7% for identifying four objects in the kitchen scenario. Besides, to measure the robustness concerning qualitative image aspects, several transformations were applied to the image. Among them are noise applications, increase and decrease of brightness and contrast and rotation about the axis and blurring applications. The algorithm's results in these transformations showed that the most significant impact factor was the noise application.

Check the feasibility to gather and build a webcam-based visual activity data set. For daily living activities in the home environment, the suggested 2D eye gaze estimation is utilized. While the webcam's quality reduces eye-tracking accuracy, the suggested webcam-based analysis technique of eye movement may be effectively used to identify everyday human activity with gaze estimation and identify the fundamental kinds of eye motions. Furthermore, accurate identification results may be demonstrated using an estimation of the eye gaze, extracting more representative internal characteristics from original variable temporal signals of eye motion. Because of its popularity and effectiveness, this model focuses on deep-learning algorithms and learning processes. Large-scale data sets and scalable computing resources, such as thousands of CPU cores and GPUs, are successfully available with profound knowledge algorithms. This research aims to improve the estimation for estimating eye gaze for elderly and disabled persons in everyday life. We aim to assist the scientific community with our work and promote better gaze-estimation methods.

REFERENCES

[1] Brandão, A., Dias, D., Alvarenga, I., Paiva, G., Trevelin, L., Gramany-Say, K. and Castellano, G., 2017, December. E-street for prevention of falls of the elderly an urban virtual environment for human–computer interaction from lower limb movements. In *Brazilian Technology Symposium* (pp. 249–256). Springer, Cham.

[2] Madhusanka, B.G.D.A. and Jayasekara, A.G.B.P., 2016, December. Design and development of adaptive vision attentive robot eye for service robot in domestic environment. In *2016 IEEE International Conference on Information and Automation for Sustainability (ICIAfS)* (pp. 1–6). IEEE.

[3] Katona, J., 2021. A Review of Human-computer interaction and virtual reality research fields in cognitive infocommunications. *Applied Sciences*, 11(6), p. 2646.

[4] Karunachandra, R.T.H.S.K., & H.M.K.K.M.B. Herath, 2020. Binocular vision-based intelligent 3-D perception for robotics application. International Journal of Scientific and Research Publications (IJSRP), 10(9), 689–696. 10.29322/ijsrp.10.09.2020.p10582.

[5] Joseph, A.W. and Murugesh, R., 2020. Potential eye tracking metrics and indicators to measure cognitive load in human-computer interaction research. *Journal of Scientific Research*, 64(1). http://dx.doi.org/10.37398/JSR.2020.640137

[6] Søgaard Neilsen, A. and Wilson, R.L., 2019. Combining e-mental health intervention development with human computer interaction (HCI) design to enhance technology-facilitated recovery for people with depression and/or anxiety conditions: An integrative literature review. *International Journal of Mental Health Nursing*, 28(1), pp. 22–39.

[7] He, Z., Chang, T., Lu, S., Ai, H., Wang, D. and Zhou, Q., 2017. Research on human-computer interaction technology of wearable devices such as augmented reality supporting grid work. *Procedia Computer Science*, 107, pp. 170–175.

[8] Appuhamy, E.J.G.S. and Madushanka, B.G.D.A., 2017. Emotion recognition and expression based on human motion in service robot eye. *Journal of Engineering and Technology of the Open University of Sri Lanka (JET-OUSL)*, 5(2).

[9] Gu, Y. and Hua, L., 2021. A novel smart motor imagery intention human-computer interaction model using extreme learning machine and EEG signals. *Frontiers in Neuroscience*, 15. 10.3389/fnins.2021.685119

[10] Knox, B.J., Lugo, R.G., Jøsok, Ø., Helkala, K. and Sütterlin, S., 2017, July. Towards a cognitive agility index: the role of metacognition in human computer interaction. In *International Conference on Human-Computer Interaction* (pp. 330–338). Springer, Cham.

[11] Triberti, S., Chirico, A., La Rocca, G. and Riva, G., 2017. Developing emotional design: Emotions as cognitive processes and their role in the design of interactive technologies. *Frontiers in Psychology*, 8, p. 1773.

[12] Jeon, M., 2017. Emotions and affect in human factors and human–computer interaction: Taxonomy, theories, approaches, and methods. In *Emotions and Affect In Human Factors and Human-computer Interaction* (pp. 3–26). Academic Press.

[13] Katona, J., 2021. A review of human–computer interaction and virtual reality research fields in cognitive infocommunications. *Applied Sciences*, 11(6), p. 2646.

[14] Santos, R.L., Abrantes, A. and Jorge, P.M., 2017. Eye gaze tracking system for adapted human-computer interface. *i-ETC: ISEL Academic Journal of Electronics Telecommunications and Computers*, 3(1), p. 1.

[15] Zheng, C. and Usagawa, T., 2018, October. A rapid webcam-based eye tracking method for human computer interaction. In *2018 International Conference on Control, Automation and Information Sciences (ICCAIS)* (pp. 133–136). IEEE.

[16] Jagadale, P.G., 2020. Role of eye tracking system to enhance life of disable people. *International Research Journal of Modernization in Engineering Technology and Science*, 2(11), pp. 715–719.

[17] Modi, N. and Singh, J., 2021. A review of various state of art eye gaze estimation techniques. *Advances in Computational Intelligence and Communication Technology*, pp. 501–510. https://doi.org/10.1007/978-981-15-1275-9_41

[18] Cáceres, E., Carrasco, M. and Ríos, S., 2018. Evaluation of an eye-pointer interaction device for human-computer interaction. *Heliyon*, 4(3), p.e00574.

[19] Meena, Y.K., Cecotti, H., Wong-Lin, K., Dutta, A. and Prasad, G., 2018. Toward optimization of gaze-controlled human–computer interaction: Application to hindi virtual keyboard for stroke patients. *IEEE Transactions on Neural Systems and Rehabilitation Engineering*, 26(4), pp. 911–922.

[20] Rajanna, V. and Hammond, T., 2018. A gaze-assisted multimodal approach to rich and accessible human-computer interaction. arXiv preprint arXiv:1803.04713.

[21] Strapper, L., Mertens, R., Pospiech, S., Bussmann, F., Grah, A. and Mamsch, M., 2017, December. A gaze tracking based, multi modal human computer interaction concept for efficient input. In *2017 IEEE International Symposium on Multimedia (ISM)* (pp. 268–273). IEEE.

[22] Parit, S.S., Dharmannavar, P.S., Bhabire, A.A., Nitave, K.N. and Patil, S.M., 2015. Eye tracking based human computer interaction. *2015 International Conference on Man and Machine Interfacing (MAMI)*. 10.1109/MAMI.2015.7456615

[23] Wang, R., Qiu, J., Luo, K., Peng, L. and Han, P., 2018, January. Eye gaze tracking based on the shape of pupil image. In *2017 International Conference on Optical Instruments and Technology: Optoelectronic Imaging/Spectroscopy and Signal Processing Technology* (Vol. 10620, p. 106201P). International Society for Optics and Photonics.

[24] Ahn, H., 2020. Non-contact real time eye gaze mapping system based on deep convolutional neural network. arXiv preprint arXiv:2009.04645.

[25] Liu, M., Li, Y.F. and Liu, H., 2020, October. Towards robust auto-calibration for head-mounted gaze tracking systems. In *2020 IEEE International Conference on Mechatronics and Automation (ICMA)* (pp. 588–593). IEEE.

[26] Li, F., Lee, C.H., Feng, S., Trappey, A. and Gilani, F., 2021, May. Prospective on eye-tracking-based studies in immersive virtual reality. In *2021 IEEE 24th International Conference on Computer Supported Cooperative Work in Design (CSCWD)* (pp. 861–866). IEEE.

[27] Zou, J., Zhang, H. and Weng, T., 2017, August. New 2D pupil and spot center positioning technology under real—Time eye tracking. In *2017 12th International Conference on Computer Science and Education (ICCSE)* (pp. 110–115). IEEE.

[28] Chen, W.X., Cui, X.Y., Zheng, J., Zhang, J.M., Chen, S. and Yao, Y.D., 2019. Gaze gestures and their applications in human-computer interaction with a head-mounted display. arXiv preprint arXiv:1910.07428.

[29] George, A., 2019. Image based eye gaze tracking and its applications. arXiv preprint arXiv:1907.04325.

[30] Luo, K., Jia, X., Xiao, H., Liu, D., Peng, L., Qiu, J. and Han, P., 2020. A new gaze estimation method based on homography transformation derived from geometric relationship. *Applied Sciences*, 10(24), p. 9079.

[31] Cáceres, E., Carrasco, M. and Ríos, S., 2018. Evaluation of an eye-pointer interaction device for human-computer interaction. *Heliyon*, 4(3), p. e00574.

[32] Majaranta, P., Räihä, K.J., Hyrskykari, A. and Špakov, O., 2019. Eye movements and human-computer interaction. In *Eye Movement Research* (pp. 971–1015). Springer, Cham.

[33] Li, B., Fu, H., Wen, D. and Lo, W., 2018. Etracker: A mobile gaze-tracking system with near-eye display based on a combined gaze-tracking algorithm. *Sensors*, 18(5), p. 1626.

[34] Chen, W.X., Cui, X.Y., Zheng, J., Zhang, J.M., Chen, S. and Yao, Y.D., 2019. Gaze gestures and their applications in human-computer interaction with a head-mounted display. arXiv preprint arXiv:1910.07428.

[35] Singh, H. and Singh, J., 2019. Object acquisition and selection in human computer interaction systems: A review. *International Journal of Intelligent Systems and Applications in Engineering*, 7(1), pp. 19–29.

[36] Mavely, A.G., Judith, J.E., Sahal, P.A. and Kuruvilla, S.A., 2017, December. Eye gaze tracking based driver monitoring system. In *2017 IEEE International Conference on Circuits and Systems (ICCS)* (pp. 364–367). IEEE.

[37] Liu, M., Li, Y.F. and Liu, H., 2020, October. Towards robust auto-calibration for head-mounted gaze tracking systems. In *2020 IEEE International Conference on Mechatronics and Automation (ICMA)* (pp. 588–593). IEEE.

[38] Zhang, X., Liu, X., Yuan, S.M. and Lin, S.F., 2017. Eye tracking based control system for natural human-computer interaction. *Computational Intelligence and Neuroscience*, 2017. https://doi.org/10.1155/2017/5739301

[39] Li, Y., Zhan, Y. and Yang, Z., 2020, June. Evaluation of appearance-based eye tracking calibration data selection. In *2020 IEEE International Conference on Artificial Intelligence and Computer Applications (ICAICA)* (pp. 222–224). IEEE.

[40] Abdel-Samei, A.G.A., Ali, A.S., Abd El-Samie, F.E. and Brisha, A.M., 2021. Efficient classification of horizontal and vertical EOG signals for human computer interaction. arXiv:2009.04645.

[41] Wang, K. and Ji, Q., 2018. 3D gaze estimation without explicit personal calibration. *Pattern Recognition*, 79, pp. 216–227.

[42] Lin, S., Liu, Y., Wang, S., Li, C. and Wang, H., 2021. A novel unified stereo stimuli based binocular eye-tracking system for accurate 3D gaze estimation. arXiv preprint arXiv:2104.12167.

[43] Ali, A. and Kim, Y.G., 2020. Deep fusion for 3D gaze estimation from natural face images using multi-stream CNNs. *IEEE Access*, 8, pp. 69212–69221.

[44] Lin, S., Liu, Y., Wang, S., Li, C. and Wang, H., 2021. A novel unified stereo stimuli based binocular eye-tracking system for accurate 3D gaze estimation. arXiv preprint arXiv:2104.12167.

[45] Santini, T., Fuhl, W. and Kasneci, E., 2017, May. Calibme: Fast and unsupervised eye tracker calibration for gaze-based pervasive human-computer interaction. In *Proceedings of the 2017 CHI conference on human factors in computing systems* (pp. 2594–2605).

[46] Mittal, M., Kaushik, R., Verma, A., Kaur, I., Goyal, L.M., Roy, S. and Kim, T.H., 2020. Image watermarking in curvelet domain using edge surface blocks. *Symmetry*, 12(5), p. 822.

[47] Yang, L., Dong, K., Ding, Y., Brighton, J., Zhan, Z. and Zhao, Y., 2021. Recognition of visual-related non-driving activities using a dual-camera monitoring system. *Pattern Recognition*, 116, p. 107955.

[48] MacInnes, J.J., Iqbal, S., Pearson, J. and Johnson, E.N., 2018. Wearable eye-tracking for research: Automated dynamic gaze mapping and accuracy/precision comparisons across devices. bioRxiv, p. 299925.

[49] Wöhle, L. and Gebhard, M., 2021. Towards robust robot control in cartesian space using an infrastructureless head-and eye-gaze interface. *Sensors*, 21(5), p. 1798.

[50] Jha, S. and Busso, C., 2020. Estimation of driver's gaze region from head position and orientation using probabilistic confidence regions. arXiv preprint arXiv:2012.12754.

[51] Chi, J.N., Xing, Y.Y., Liu, L.N., Gou, W.W. and Zhang, G.S., 2017. Calibration method for 3D gaze tracking systems.*Applied Optics*, 56(5), pp. 1536–1541.

[52] Han, S.Y. and Cho, N.I., 2021, January. User-independent gaze estimation by extracting pupil parameter and its mapping to the gaze angle. In *2020 25th International Conference on Pattern Recognition (ICPR)* (pp. 1993–2000). IEEE.

[53] Adithya, P.S., et al. 2019. Design and Development of Automatic Cleaning and Mopping Robot, IOP Conference Series: Materials Science and Engineering, vol. 577. no. 1. IOP Publishing

[54] Wöhle, L. and Gebhard, M., 2021. Towards robust robot control in cartesian space using an infrastructureless head-and eye-gaze interface. *Sensors*, 21(5), p. 1798.

[55] Lindén, E., 2021. Calibration in deep-learning eye tracking. Doctoral dissertation, KTH Royal Institute of Technology.

[56] Emery, K.J., Zannoli, M., Xiao, L., Warren, J. and Talathi, S.S., 2021, March. Estimating gaze from head and hand pose and scene images for open-ended exploration in VR Environments. In *2021 IEEE Conference on Virtual Reality and 3D User Interfaces Abstracts and Workshops (VRW)* (pp. 554–555). IEEE.

[57] Velisar, A. and Shanidze, N., 2021, May. Noise in the machine: Sources of physical and computation error in eye tracking with pupil core wearable eye tracker: Wearable eye tracker noise in natural motion experiments. In *ACM Symposium on Eye Tracking Research and Applications* (pp. 1–3).

[58] Lu, Y., Wang, Y., Xin, Y., Wu, D. and Lu, G., 2021, June. Unsupervised gaze: Exploration of geometric constraints for 3D gaze estimation. In *International Conference on Multimedia Modeling* (pp. 121–133). Springer, Cham.

[59] González-Ortega, D., Díaz-Pernas, F.J., Martínez-Zarzuela, M. and Antón-Rodríguez, M., 2021. Comparative analysis of kinect-based and oculus-based gaze region estimation methods in a driving simulator. *Sensors*, 21(1), p. 26.

[60] Yeamkuan, S. and Chamnongthai, K., 2021. 3D point-of-intention determination using a multimodal fusion of hand pointing and eye gaze for a 3D display. *Sensors*, 21(4), p. 1155.

[61] Mittal, M., Arora, M., Pandey, T. and Goyal, L.M., 2020. Image segmentation using deep learning techniques in medical images. In *Advancement of Machine Intelligence in Interactive Medical Image Analysis* (pp. 41–63). Springer: Singapore.

[62] Sanjeewa, E.D.G., Herath, K.K.L., Madhusanka, B.G.D.A. and Priyankara, H.D.N.S., 2020. Visual attention model for mobile robot navigation in domestic environment. *GSJ*, 8(7). Online: ISSN 2320-9186

[63] Herath, K.K.L., Sanjeewa, E.D.G., Madhusanka, B.G.D.A. and Priyankara, H.D.N.S., 2020. Hand gesture command to understanding of human-robot interaction. *GSJ*, 8(7). Online: ISSN 2320-9186

[64] Abeyrathne, W.S.L., Madushanka, B.G.D.A. and Priyankara, H.D.N.S., 2020. Vision-based fallen identification and hazardous access warning system of elderly people to improve well-being. *International Journal of Engineering Research & Technology (IJERT)*, 9(8).

[65] Moladande, M.W.C.N. and Madhusanka, B.G.D.A., 2019, March. Implicit intention and activity recognition of a human using neural networks for a service robot eye. In *2019 International Research Conference on Smart Computing and Systems Engineering (SCSE)* (pp. 38–43). IEEE.

[66] Vithanawasam, T.M.W. and Madhusanka, B.G.D.A., 2018, June. Dynamic face and upper-body emotion recognition for service robots. In *2018 IEEE/ACIS 17th International Conference on Computer and Information Science (ICIS)* (pp. 428–432). IEEE.

[67] Milinda, H.G.T. and Madhusanka, B.G.D.A., 2017, May. Mud and dirt separation method for floor cleaning robot. In *2017 Moratuwa Engineering Research Conference (MERCon)* (pp. 316–320). IEEE.

[68] Vithanawasam, T.M.W. and Madhusanka, B.G.D.A., 2019, March. Face and upper-body emotion recognition using service robot's eyes in a domestic environment. In *2019 International Research Conference on Smart Computing and Systems Engineering (SCSE)* (pp. 44–50). IEEE.

[69] Madhusanka, B.G.D.A. and Ramadass, S., 2021. Implicit Intention communication for activities of daily living of elder/disabled people to improve well-being. In *IoT in Healthcare and Ambient Assisted Living* (pp. 325–342). Springer, Singapore.

[70] Madhusanka, B.G.D.A. and Sureswaran, R., 2020.Understanding activities of daily living of elder/disabled people using visual behavior in social interaction. Asia Pacific Advanced Network (APAN50). Hong Kong.

[71] Appuhamy, E.J.G.S. and Madhusanka, B.G.D.A., 2018, June. Development of a GPU-based human emotion recognition robot eye for service robot by using convolutional neural network. In *2018 IEEE/ACIS 17th International Conference on Computer and Information Science (ICIS)* (pp. 433–438). IEEE.

[72] Hassoumi, A., Peysakhovich, V. and Hurter, C., 2019. Improving eye-tracking calibration accuracy using symbolic regression. *PLOS One*, 14(3), p. e0213675.

[73] H.M.K.K.M.B. Herath, Karunasena, G.M.K.B. and Herath, H.M.W.T., 2021. Development of an IoT based systems to mitigate the impact of COVID-19 pandemic in smart cities. In *Machine Intelligence and Data Analytics for Sustainable Future Smart Cities* (pp. 287–309). Springer, Cham.

[74] H.M.K.K.M.B. Herath, 2021. Internet of things (IoT) enable designs for identify and control the COVID-19 pandemic. In *Artificial Intelligence for COVID-19* (pp. 1–14). Springer, Cham.

[75] Mittal, M., Verma, A., Kaur, I., Kaur, B., Sharma, M., Goyal, L.M., Roy, S. and Kim, T.H., 2019. An efficient edge detection approach to provide better edge connectivity for image analysis. *IEEE Access*, 7, pp. 33240–33255.

[76] Liaqat, A., Khan, M.A., Sharif, M., Mittal, M., Saba, T., Manic, K.S. and Al Attar, F.N., 2020. Gastric tract infections detection and classification from wireless capsule endoscopy using computer vision techniques: A review. *Current Medical Imaging*, 16(10), pp. 1229–1242.

[77] Dash, S., Acharya, B.R., Mittal, M., Abraham, A. and Kelemen, A.G. eds., 2020. *Deep Learning Techniques for Biomedical and Health Informatics*. Cham: Springer.

Index

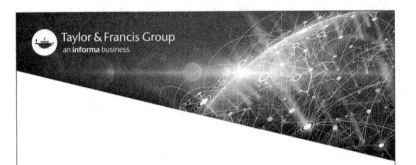

Printed in the United States
by Baker & Taylor Publisher Services